RECOVERING A LOST RIVER

Snake River Watershed

1:7,250,000

Albers Equal Area Projection

0   37.5 75   150   225
Miles

Niels Maumenee

RECOVERING A LOST RIVER

Removing Dams,
Rewilding Salmon,
Revitalizing Communities

Steven Hawley

BEACON PRESS ~ BOSTON

BEACON PRESS
25 Beacon Street
Boston, Massachusetts 02108-2892
www.beacon.org

Beacon Press books
are published under the auspices of
the Unitarian Universalist Association of Congregations.

14 13 12 11   8 7 6 5 4 3 2 1

This book is printed on acid-free paper that meets the uncoated paper
ANSI/NISO specifications for permanence as revised in 1992.

Text design and composition by Wilsted & Taylor Publishing Services

Library of Congress Cataloging-in-Publication Data

Hawley, Steven.
  Recovering a lost river : removing dams, rewilding salmon, revitalizing
communities / Steven Hawley.
    p. cm.
  Includes bibliographical references.
  ISBN 978-0-8070-0471-5 (hardcover : alk. paper)
    1. Stream conservation—United States. 2. Stream restoration—United
States. 3. Watershed management—United States. I. Title.
  QH104.H39 2011
  333.91'621530973—dc22   2010034523

For Elliot and Annabel

Expect poison from the standing water.

—WILLIAM BLAKE,
The Marriage of Heaven and Hell

CONTENTS

PROLOGUE

The seeds for this book were fertilized years ago, when I poisoned a dozen people on a float trip through Hell's Canyon. Missoula, Montana, my home then, was less than a day's drive from some of the world's greatest rivers. So great in fact, that they require you win the lottery to get on any one of them—the United States Forest Service lottery, which determines which lucky contestants get to launch on the Middle Salmon, Main Salmon, Selway, and Snake rivers. Although the Snake, or rather the last free-flowing portion of it through Hell's Canyon, is considered something of a consolation prize in this four-river draw, any chance to float through these wild places is usually taken by the free-thinking and free-spirited—come hell or high water.

This scenario creates an unusual, perhaps unprecedented circumstance in modern-day travel: the lottery winner, upon whom luck has smiled, often feels obliged, or perhaps is coerced, into sharing her good fortune by inviting a few friends, who in turn may invite a few more friends. Before long, a loose affiliation of folks whose backcountry capabilities, quirks in personality, peccadilloes in daily habits of living, or prior arrest records are unknown to one another will embark on a week-long adventure through a stretch of country where cell-phone reception remains a space-age fantasy.

This was such a trip. We loaded a convoy of cars with rafts, tents, sleeping bags, sleeping pads, guitars, kayaks, a solar shower kit, a first-aid kit, paddles, oars, Frisbees, a Wiffle ball and bat, coolers, folding tables, folding chairs, a portable privy, fishing rods, life jackets, helmets, a full-service kitchen and bar, and an item that later proved to be indispensable, a hand-cranked stainless-steel blender.

Right away, our group showed great courage and self-discipline: it was the zenith of summer's heat, late July, and no one even mentioned getting into the coolers of beer until we were almost a third of the way to the put-in. Already, we had traveled a long way: from Missoula, west down twisted Highway 12, following the Lochsa River to its confluence with the Selway, where the Clearwater commences, then on to Kooskia, Idaho, where the South Fork of the Clearwater meets the main stem. We skipped rocks in the river where these two currents meet, then proceeded on: up the South Fork, up the Harpster Grade to Grangeville, then over Whitebird Pass, and on down into Salmon River country; the Seven Devils on our right, the Salmon-Selway wilderness on our left. In a mood of anticipatory delight, we stopped to swim at a white-sand beach adjacent the river, then rested in Riggins.

We then moved up the road along the Little Salmon River: eventually on to Council, then Cambridge, up the grade called Brownlee Summit, and down once more to Oregon and the fetid slack water behind Oxbow Dam on the Snake River. The end of the day's long drive, a cursory tour of the edges of the largest tract of wilderness in the Lower 48, proved to be something of a disappointment: here lay what used to be a wild river, the upper portions of Hell's Canyon.

At the well-manicured turf-and-concrete campground below the dam, graciously provided by Idaho Power as reparation for the missing river, we had finally arrived. We had missed the dinner hour by a long shot, and a hungry crew was grumbling in the gloaming about getting some food down their gullets and getting to bed. Almost everyone was there, except the guy with the full-service kitchen and bar in the back of his truck.

I alone faced a minor public health dilemma. For dinner that night, I had prepared in advance a simple Mexican dish, chicken mole. The custom was simply to buy a whole chicken, boil it for an hour, place it on a platter, prepare the mole sauce, place that along with the salsa, cheese, onions, sour cream, and other trimmings in buffet-style bowls, and let dinner guests create their own sweet burritos. That was the usual procedure, but in the rush to pack everything in, I found it necessary to take some shortcuts that proved to be costly. In an effort to save cooler space, I carved the bird in advance, placing the strips of meat in the sauce, sealing the whole affair into a giant plastic container. Arriving upon the chaos of the group packing, I discovered all the coolers were full: no matter, I would shuttle home, grab another cooler, put the evening's repast on ice, and all would be well.

"Don't bother, we've got seven coolers already," someone had said.

Regrettably, I didn't act on this simple leap in rational thought: seven coolers, all full to the hilt, so an eighth would be required on a ninety-eight-degree day to get dinner safely to its destination. I shrugged, capitulated to the notion that seven coolers were enough, and the plastic container traveled uncooled for the duration. I noted with silent concern on several occasions over the journey the signals of a volatile chemical reaction. As I glanced furtively through the transparent plastic, the contents therein appeared to be boiling still, just as they had been when transferred that morning.

As darkness descended on our starving contingent, I made an executive decision: we could not eat until the chicken was heated. Eyes rolled. Several people stomped off in the dark. One woman was moved to tears. "So where's the stove?" someone asked.

"It's in Gary's truck."

"Where the fuck is Gary?"

"Who the fuck is Gary?"

"I think they went to the put-in."

"That's seventeen miles from here."

"Great. How about a little drive?"

Once again, we proceeded on, sending one car speeding

ahead. After what seemed like an hour, we met our pilot car coming back the other way.

"I couldn't find'm, man, no sign."

"Should we drive back?"

"We should just eat."

I felt the eyes of a half-dozen perfect strangers burning into me. I popped the lid on the chocolate chicken. It smelled fine. I dipped my finger in the sauce. It tasted even better.

"Screw them. Let's eat," I said.

With blazing efficiency, two tables were set up on the roadside, dinner was laid out, and a gaggle of happy campers were stuffing their faces. Moments later, Gary pulled up, headlights blinding.

"We saw your car, dude, you drove right past us," he explained. "We were at the put-in, wondering where everyone was, and saw you and figured we'd better backtrack."

"Thanks for sticking with us," someone mumbled sarcastically in the darkness.

An awkward silence ensued.

"Wow. Cold chicken burritos," Gary observed.

"Not exactly cold."

"Not exactly hot."

"Shut up, whiners. The beer's cold."

Mellowed with full bellies, we decided there was enough room for all of us to camp right where we stood, at a wide spot in the road, on a concrete slab near the edge of a reservoir that looked like it had been used for staging equipment during the dam's construction. One couple argued about how to set up a tent on cement. Another griped about the "eight goddamn hours" it had taken to get here. An ominous tension cast a pall over the evening.

I decided a solitary stroll was in order. I tried to let the complications of the group dynamics roll off me like water off a duck's back. Thoughts drifted along in the night, sluggish as the river's current between two dams.

It was the dam's fault, I concluded. We should be camped

along the grassy banks of a free-flowing river, more like the spot on the Salmon where we'd swum earlier in the day. Instead, we were parked at an industrial ruin: run aground on cement overlooking the seventy-two-degree remains of one of the greatest rivers on the planet. Bass and squawfish where there should have been steelhead and salmon. Power lines and riprap where there should have been giant ponderosas and a silvery moonlit stream studded with giant boulders. I watched the distorted reflection of the moon twist and stretch on the surface of the water. The night sky was framed by the canyon walls, a vestige of the river's former glory. I tried to use these images as a defense against a toxic, all-consuming despair, a pointless rage I've too often felt in my life at times and places where something marvelous and alive has been subverted, submerged, capped in concrete, and memorialized with a plaque.

I imagined the river at work. It had never really quit; it was there now, pressing on, working in the microscopic fissures and cracks of the aging and weakening concrete. Eventually, the river would once again find a way through. Maybe not in my lifetime, but in its own time, it would. Having found a modicum of solace, I turned back to the cement slab, found my spot, unrolled my sleeping bag, and slipped off to sleep.

It turned out that the toxic feeling I'd fought that night was not entirely a result of the vagaries of the modern age. I had been dreaming, as I reliably do when bedded down with dark musings, of a bleak, smoke-filled future, hemmed in by wrought iron fences, protected from some airborne menace by steel netting, harassed by ragged, starving children armed with assault rifles, watched over from a guard tower by some armor-plated, omnipresent, coldly militaristic overlord, not a flower or river or even the sun anywhere in sight.

I walked the terrain of my recurring nightmare in vivid terror until I sat bolt upright in my sleeping bag. Or at least thought I did. I must have been dreaming still. In what other circumstance would strangers be running about half-naked on crumbling concrete in pitch darkness, stumbling like the demented tenants of a dim sanitarium, rolls of toilet paper clutched desperately in their hands? In my dream, I laid back down to return to sleep. I heard tortured moaning, then what sounded like a horse fart. Suddenly, I realized I was awake, and that if I did not move with extreme caution, but also with utmost urgency, I was going to have the kind of accident that I hadn't had since not quite making it home in time from Widgeon League baseball practice when I was five.

After negotiating the gymnastics pertaining to the mere inaugural rounds of my emergency, I inquired into the general status of the group. Not good: the road on either side was a rocky, basalt-strewn talus slope, with a dearth of cover for a dozen violently ill people to seek a quick and somewhat private shelter. A few scrubby sagebrush were the only hope. Eyeballs peered out in distress from every possible hiding place; squatting naked haunches appeared as desperate apparitions in the night, bringing to mind the feckless, doomed plight of a herd of diseased ungulates. Another contingent had already driven off on a high-speed shuttle to the outhouse near the spillway of the dam, but that one-holer could not provide much help for the five folks on their way there.

My friend John approached, looking none the worse for wear.

"Everyone's sick except me. I'm on antibiotics," he serenely confessed without a twinge of survivor's guilt. "And Gary, he's got a super-crazy metabolism. He got it first, but he seems to think he's done with it."

Gary came bounding toward us in the dark, headlamp trained on us like a rescue beacon, energized and victorious as only a man who's licked a sickness can be. His dark eyes shone; his wild, curly black hair stood on end.

"Hey dude!" He slapped us each a high five, his customary greeting. "Just about shit the bed about 2:30. Got up and hung it right into the water," he said matter-of-factly, thumbing toward the reservoir.

"Excuse me," I said, sprinting in that direction.

"Don't worry, dude!" Gary yelled after me. "I bet by the time we put on, everyone'll by fine. Let 'er rip! Holy Molaay!"

Gary, it turned out, was right—everyone was more or less fine by noon that day. Everyone except me. The chicken meted out its worst vengeance on the cook, and the rest of the group, I suppose, took a deserved pleasure in my pain. We cleaned up our camp—as best we could anyway—and made our way to the visitors station at the base of the dam adjacent the put-in. While the rest of the crew endured a lecture from a Forest Service ranger (fire in the pan, solid waste in the rocket box, no fireworks, don't pick the flowers, watch out for bears, snakes, goats, sheep, lions, and poison oak) and commenced preparation of the flotilla, I endured the pungent, par-boiled stench of the outhouses near the boat launch. When I emerged a few hours later, everything was set. Spirits were high, all crafts were manned or womanned, and a fourteen-foot blue raft assigned to my good keeping awaited my somewhat tender keister in the rowing seat. Mercifully, it was an uneventful day on the water.

That night, we all went to bed early and slept the sleep of angels, snoozing well past sunrise. We were greeted with electric blue skies and enjoyed a lethargic, indolent morning. The day was flawless, slow, idyllic, unhurried, filled with carefree laughing, swimming, fishing, drinking, eating, joking, and playing. We set up camp an undetermined distance downstream—no one cared how far we'd come or how far we had to go—beneath a basalt cliff with a sheltered alcove at water's edge. This served as the kitchen area, once the rattlesnake that occupied it was evicted for the evening. Not far upstream, a hundred paces or so, a creek, cold, clear, and running strong through the thickest of brush, provided fresh water, and for me at least, enticed a walk to its source, nested in a high lake or spring in the canyon's rim. A day's walk, albeit through thickets of poison oak that made me itch just thinking of them. I settled for a sponging off and a long soaking of the feet as I pumped drinking water for the crew.

Along came Carter, an eccentric individual I had seen but

never met back in Missoula. Seen, because Carter drove a beat-up pickup, camouflaged in convincing fashion in the manner of the fish represented in the genus oncorhyncus: silvery flash on bottom, blurring into steel gray on top, sporting speckles along the body, and orange slashes along the grill where the signature marking of a cutthroat would be. Another Missoula trout bum.

"Holy Molay," Carter coolly greeted me. "See any fish?"

"No fish," I said.

"Can you imagine this river before all the dams were in?"

"Yes," I said. "I think I can."

"Me too," said Carter, with a fisherman's faith in things unseen.

Just then, Gary appeared, carrying what appeared to be two miraculous frozen margaritas.

"Here guys. We were gonna save this for the last night, but it's been too good a day."

He handed over the drinks and received a hero's praise. With the first sip, I could feel the balm of ice, lime, Triple Sec, and tequila repairing and actually strengthening my gastro-intestinal tract, a warm glow emanating from the belly that signaled a full recovery was underway.

"What d'you think about tearing out dams?" Carter asked Gary, the tequila apparently prompting him to persist in his line of dam questions.

"Tear them all out, I'd say."

"You think?"

"I'll tell you what I think, I think it's a great idea," proclaimed Gary. "It would be sweet. You could put on Marsh Creek like we did this spring and run the whole stretch. The Middle, the Main to the Snake at Heller Bar, and then the Snake all the way to Pasco. All the way to the confluence with the Columbia. The trip of a lifetime.

"Just as good as a month on the Grand Canyon, no problem. Plus you could fish the whole way," he finished, slapping us all high fives at the idea.

The dinner bell rang shortly thereafter. Someone had

prepared a safe but delectable pasta dish that night. The wine flowed freely. The hand-cranked blender kept cranking. The soft gold, red, then blue light in the canyon ushered the stars out, and a hot evening breeze wound down to pleasantly warm. The whole crew hiked the hundred paces up to that creek in the dark, all of us tipsy, some barefoot, no lights, no lifejackets, across a loose, rocky river bank, and plunged into the river and floated back to camp. A foolish lark, poor judgment—somebody could have broken an ankle, cracked a skull, gulped a lungful of river water, or spilled a drink. Yet so absolutely necessary, we decided it should be done once more. Boisterous, reckless laughter followed, then a few more hours of quiet, intimate conversations beneath the starlight, and then, for everyone but me, eight to ten hours of blissful deep sleep, with nothing but the whisperings of the river to interrupt.

I love to sleep outside. Love it so much, in fact, I sometimes spend a good portion of the night awake. Like many soft, suburban-raised nature nuts, I have a powerful fear of snakes, scorpions, fire ants, pissants, raccoons, bears, pack rats, gnats, ticks, and fleas, the rustling in the brush, the glowing marble eyeballs rolling low over the ground in the beam of a flashlight. Unlike many of my compatriots, I won't set up a tent if I feel I can get away with it. The chance to sleep out under the stars, to see clouds scudding across the moon, to feel dew falling on my face, to be awake at night and stare down, between the narcotic throes of dreams, the strange and mysterious darkness, is worth the interruptions of those precious REM hours for me. And so it was that evening, lying on my back along the banks of the Snake, meditating on what the river used to be, what it was at that moment as I lay there half awake, and what it might become, even as I came ever closer to finally shutting my eyes for the night. Suddenly, I detected the presence of a little intruder in my bed, crossing into territory familiar only to a lover or attending physician.

With the reflexes of a ninja, I sprang out of bed, shook the unwelcome visitor out of my bag, then found my way by the light of the moon to the kitchen, where the dregs of the last pitcher from the hand-cranked blender had melted. I poured what was left in my trusty tin cup, and set off on a moon walk up river, where I was drawn once again to the creek. I sat at the confluence of the little creek and the big river, enjoying the sweet, cool water on my feet, when I noticed a swirling on the surface of the first pool in the creek not ten feet away. Two silhouettes, each one six or eight inches long. A pair of young trout or salmon smolts, seeking refuge from the overheated main stem, undulating patiently with the current, probably waiting for the river to cool a few degrees. Waiting until when? Higher up, in a week or two, the nights would begin to chill precipitously. Down in the brackish lowland heat of Hell's Canyon, it could be a month or longer before the river dipped back below seventy degrees. Not much of a summer vacation. Succumbing to the acutely goofy but somehow obligatory compulsion to make some gesture of good will, I bowed at the pool, then raised my outstretched arms to the moonlit sky, tried to stifle a yawn in my moment of reverence, failed, and so returned to my sleeping quarters, where the rest of the night passed in uneventful unconsciousness.

I thought of those fish quite often while writing this book. I wonder what became of them, just as often as I think of the good people with whom I spent those days on the river, some of whom, like the fish, I haven't seen again.

I recall this episode at the outset, with a nod to the idea that a writer owes a reader an introduction, some revealing sampler of belief, preference, or personality before setting forth together on any sort of narrative adventure. That bad chicken might be overcome by good company on a good river is not precisely the point. Nonetheless, the setting and the themes of that trip those years ago coincide with those found in this book: a restoration of good health and a variety of redemption to be found in and near flowing water.

Water, it has been said, forgives us again and again. That it does so is as much a hydrological reality as a hopeful metaphor.

How we might reciprocate for such an eternal fountain of forgiveness is one of the great philosophical questions water raises. If, as Thoreau wrote, "The life in us is like the water in the river," it should be clear that we have been courting trouble for a while now. A third of freshwater species are threatened or endangered. Forty percent of freshwater bodies in America are too polluted for swimming or fishing. Half the nation's wetlands are gone. The oceanward progress of six hundred thousand linear miles of brook, creek, stream, and river has been retarded by the construction of 75,000 dams. Water may forgive our trespasses, yet we continue to trespass against it.

Like many of the more practiced ecological sins of the age, this profligate wasting and abuse of water, we've been warned, will have to come to a stop sometime in the first half of this century, lest a critical mass of transgression morph into an unpardonable sin. This book examines one possible route to averting that catastrophe. The journey begins with a quality humans passionately have come to admire in the peregrinations of the animals: loyalty to home. This is the much celebrated sense of place perhaps overly admired in so much outdoor literature. It may make for pretty prose, but it comes with baggage.

For instance, the creeks and rivers around my current home, roughly defined as the southern half of the Columbia River watershed, have infected me with a virulent strain of biophilia, to borrow E. O. Wilson's apt phrase. I've felt that I am being spoken to by these great bodies of mountain water and have even experienced the fleeting glimpse of a sensation that I might speak back to them. But I'm also aware that these rivers are fast transforming me into something of a provincial imp. No matter where else I travel on the globe, beyond any river that flows west to the Pacific, amid scenes of unforgettable splendor, I cannot help but feel pity or even contempt for these lesser rivers. They are flowing the wrong way, not exactly wasted but somehow not adequate, something akin to the crestfallen mood that strikes when observing a bear eating garbage. Natural, to be sure, but not part of a better world, a more perfect order of things.

In addition to this kind of narrow-minded provincialism, loyalty to home breeds other kinds of trouble. A deep love of a particular landscape has become a prerequisite for experiencing the punishments of the star-crossed, the condemnations of the perennial loser. Fighting for a favored trout stream or flower meadow has mostly been a losing battle, if only because the defenders, usually outgunned and outmanned, only have to lose once.

Yet in the face of the litany of ecological and social disasters facing mankind at the beginning of the twenty-first century, a growing number of defenders of local creeks, grasslands, and birds are sticking to an increasingly viable notion that centering the recovery of human health on the restoration of wild animals and wild places might make the unpleasant task of reckoning with life in an era of limits not only more tolerable but turn us with eagerness toward the pleasures of undoing harm. "Hope is the thing with feathers," observed Emily Dickinson; if this is possible, then it's not too much of a stretch to proclaim it is a thing with fins as well.

One discipline within this art of replenishment, river restoration, operates as a democratizing and occasionally galvanizing force in local communities. Serendipitously pleasant sensations seem to accompany gestures toward improving local water. As one patriotic old farmer in the Columbia Basin said it, working alongside people he would not, on general principle, be seen fraternizing with in town, "These river restoration folks are bringing people together in this country in a way I haven't seen since World War II." Somewhat counterintuitively, this book will examine the notion that blowing up dams might be the environmental movement's equivalent of bombing Rommel's railroad bridge in Tunisia in 1942: an offensive, rather than defensive, position, a surgical strike that might turn the momentum of a war.

Unexpectedly, Americans have quickly gotten into the swing of destroying dams. The inventory of dams in the country has been reduced by nearly five hundred in the last decade. Most have come out on smaller, privately owned projects whose owners realized the cost of complying with modern safety standards and popular environmental laws eclipsed the benefit of keeping title to the concrete in question.

To the dismay of defenders of dams, there does seem to be an element of truth to the domino theory of dam removal. Numerous smaller projects have led to proposals for bigger ones. The Holy Grail is the plan to remove four large dams on the lower Snake in Eastern Washington. Of more than fifty studies on the subject, none have found it to be a terrible idea. All but a half dozen concluded it's both economically feasible and scientifically well-warranted. Removing these dams may be the only way to fully restore endangered salmon runs, as several federal laws and treaties mandate. It would open up unfettered access to the formerly rich salmon-bearing rivers draining more than three million acres of federal wilderness.

However justified ecologically, the most difficult obstacle toward making the Snake dams disappear is political. Yet loyalties on this front are shifting as well. Advocates of ambitious, large-scale ecological restoration projects are often perceived as the fringe element of the environmental movement. Tenured biology professors, professional conservationists, part-time yoga instructors, and feckless young trustafarians might compose the stereotypical constituency of pipe-dream projects bringing back the Buffalo Commons or designating a wildlife corridor from the Yukon to Yellowstone. But the idea of restoring salmon to the Snake has begun to engage citizens from a much broader social spectrum. The fanfare for dam deconstruction comes on the heels of a growing realization that dams do not provide an environmentally benign, economically beneficial source of electricity, as dam defenders often claim.

Slowly changing attitudes in Idaho and Eastern Washington come on the heels of some unfulfilled promises made by the federal agencies that pitched for ever more federal dams at the height of the Cold War. In addition to swearing an oath not to inflict permanent damage to the Snake's salmon, the United States Army Corps of Engineers sold these dams as tools of economic prosperity and national security. The dams have struck out in all three of these areas, giving a lot of people who don't think of themselves as dedicated enviros reason to reconsider what has been an unquestioning loyalty to the values symbolized by big federal dams.

Defending the dam status quo on the Columbia and Snake in the face of these shifting attitudes has been accomplished by a cabal of government agencies involved in managing the basin's salmon. They refer to themselves collectively as the "federal family." The family is obliged by law and treaty to protect and restore salmon. But their actions too often have worked like a protection racket the mob might offer residents of a bad neighborhood. Thus a more figurative but ultimately just as ecologically damaging form of obstruction to free-flowing waters has arrived in the form of the long shadow cast over water policy by an all-too-familiar partnership between industry and government. Some understanding of this undue influence, described in the pages that follow, is key to comprehending the discussion over Snake River dam removal.

In talking to many fisheries biologists, a curious professional colloquialism often made its way into my notes. If a chosen course of action seemed to be showing some benefit for the fish, these scientists would tell me, "The fish are talking." I came to admire this turn of phrase. If the fish are talking, it offers us an outside chance that they'll stick with us a little longer, if only we can show them we hear. What follows, then, is for all the eyes and ears on the water.

RECOVERING A LOST RIVER

1

REDEEMING THE DAMMED

*It is not clear at this time who is entitled to an accolade for
dynamiting the south abutment. However, there is no limit to what
can be accomplished if no one cares who gets the credit.*

—JOHNNY CAREY AND CORT CONLEY, *River of No Return*

On a gin-clear June afternoon, the high granite escarpments
of Idaho's Sawtooth Range are awash in brilliant sunlight. The
sky shimmers a deep electric blue; the air, light and sweet, is
laden with the smell of pines, new grass, and mountain flowers.
A breeze brings an intermittent hint of the high-lonesome scent
of bittersweet sage. In this broad subalpine valley known as the
Stanley Basin, some person or persons unknown blew up a dam
on behalf of a river's namesake fish.

Not far from here in 1910, the Golden Sunbeam Mining
Company staked a claim up the Yankee Fork of the Salmon
River. To supply power to the mine, they plugged the main river with three stories of concrete that created an impenetrable
barrier for sockeye, steelhead, and chinook salmon swimming
to their home waters more than a mile above sea level, better
than a thousand miles from the ocean. Officially, the story of
how Sunbeam Dam was destroyed remains one of Idaho's more
compelling historical mysteries. The causes for local grievance
were obvious enough. The gold mine had gone belly-up in less
than a decade. The venture capitalists who gambled on hitting the mother lode skipped town with what was left of some
Easterners' seed money and, as was customary, more than a few

1

broken promises. The dam stayed long after the hucksters who claimed to have a good reason for building it were gone. Then one morning in 1934, the dam was gone, too.

Where the Salmon River begins to narrow into its first deep canyon as it drains the eastern side of the Sawtooths and Stanley Basin, you can pull off Highway 93, scramble down a steep bank, and walk 250 feet out on the crumbling crest of Sunbeam, stopping (hopefully) where its original length was tailored to meet the needs of the local fishing community. The well-planned demolition work created a healthy rapid. Standing 30 feet above the froth, you can feel the hydraulic tremors of the standing wave beat out a rhythm in your rib cage.

If you look at the dam from a safer vantage point some forty yards downstream, its bare concrete backside strung across the river like a blank canvas, the scene leaves the impression of a huge landscape painting whose creator did such a convincing job rendering the left quarter of the scene, it came alive and sucked him into the river before he could finish. Along these lines, it would seem that Sunbeam has evolved over the years into a work of art. Certainly, the act of destroying the dam was not intended to be that. But like any relic that survives the weather, wars, and the inexorable creep of progress for the requisite period of time, an artful rendering of the ruin arises. And like any worthwhile accidental art, the questions surrounding Sunbeam tend to turn conventional wisdom on its head. Its short unhappy life, for starters, suggests that its construction may have been a greater crime than its destruction.

The art of resistance, then, becomes one of Sunbeam's prevalent themes as well. Getting rid of a fish-killing dam, or any other form of social menace, is an exercise in a variety of justice that's served when ordinary citizens act on their finer instincts. Sunbeam, or what is left of it, stands as a monument to the idea that achieving some form of communal good may require a departure from mere good behavior. Restoring the seeds of harmonious creation might lie in a singular act of destruction, an idea that must be executed artfully if it is to avoid degenerating into the senseless violence of terrorism or petty meaningless vandalism.

A leisurely exploration of Sunbeam might as well invite meditations on absence and presence, a trope familiar to art farts and nature freaks alike. It is the missing piece of the dam that renders it obsolete, just as the presence of a completed dam obliterates the existence of a working whole, an intact river system. The salmon that inspired Sunbeam's partial disappearing act are themselves mostly vanished these days, just a relative few token fish left each year after eons of plenitude. Where the skeleton of a dam and the ghost of a river collide, Sunbeam might be construed as the symbolic gateway to a river still caught between the devil and the deep blue sea.

Perhaps the most intriguing aspect of Sunbeam remains its standing invitation to imagine what really happened back in 1934. Not yet corralled within the fence of widely corroborated historical fact, the observer in this case is free to recreate just what peculiar mix of personalities and circumstances inspired an act that presaged by more than a half century the advent of a broader effort to free the nation's rivers from their dams.

The scene back then wouldn't have been too much different than today: mountains ringing a cerulean dome of sky; the river rolling oceanward, humming a tune from the vast canon of music created by water moving over rock. Before sunrise, we see a man with calloused fingers, hardened by years of work in nearby mountain mines, confident in his ability to execute the task before him, steadily, stealthily laying flame to a three-minute length of fuse. He and two others have worked through the night, making their way after dark down into the bowels of the dam. The night watchman, sympathetic to their cause, looks the other way. Working with a star bit and sledgehammer, two of the men set about drilling holes and setting charges. Commencing their labor with a familiar rhythm, one hammers the bit while the other twists the foot-long, threaded metal spike half a rotation after each metallic strike. They make easy progress into the cement of the south abutment, increasing their confidence that the dam could be separated from the relatively soft rhyolitic rock to which it was anchored. The third man keeps watch through the night atop the dam and, well past midnight, carefully feeds the contents of four boxes of TNT, along with the requisite blasting caps and wires, down to his compatriots in the tunnel below. At the first hint of dawn, the watchman and two others retreat, piling in a car and following the road downriver toward Challis. Fuse lit a short while thereafter, the last man does not run but saunters toward a carefully chosen safe haven.

Two minutes and forty-five seconds into the interval the fuse provides, a belted kingfisher sounds her alarm call and steals away, flying over the dam downriver, the branch in an old snag where she was perched just above the south abutment wobbling like a diving board after a dive. The first charge detonates. In a split second, the sway of that single branch appears to spread like a shiver and then a sneeze that knocks the snag into the water as the shockwave rushes past, the compression of air rippling outward, registering a low rumble as far away as Galena Pass. A few moments later, the river awakens to the memory of its full current. The old snag begins to float toward the cloud of still-settling dust, a fine powder of debris that falls to the river, sounding like rain. The buoyant dead tree begins to rotate slowly in the accelerating flow. The cloud downstream dissipates. Where a permanent-looking obstruction had been just a few moments before now appears a horizon line, the lip of a precipice. The tree enters the lead-in to a rapid no one has ever seen before. Parallel to the current, for an instant, it seems perfectly positioned to make a run clean through, but at the last second, it gets pushed sideways, resisting the ride downstream, slowing as if in protest of the pending collision. The butt of the snag, rotten rootwad and all, makes an audible thump as it smacks against the remaining north section of the dam, while the upper half of the tree from which the kingfisher so recently fished for breakfast is violently pummeled in the froth. The lone man watches and listens to the few seconds of feeble grinding and scraping before the snag squirts like a watermelon seed between thumb and forefinger into the thalweg of reinvigorated current. Satisfied with his small part in returning some free-flowing water to the river known as the River of No Return, he recedes like a shadow at twilight back toward Stanley and the mountains beyond.

~ ~ ~

The skilled demolition work that freed the Salmon River back then has proven to possess a long, well-amplified echo, the artistic flair grown evermore brazen. In 2007, on the other end of the same vast Columbia River watershed, outside Portland, Oregon, it was confirmed that four thousand pounds of dynamite expertly arranged to blow up a working hydroelectric dam can draw a lot of attention. The perpetrators in this instance knew something about how to play on the media's penchant for sensationalism. On a hazy summer morning, the ignition wires they'd laid were bundled to an old-fashioned detonation plunger for dramatic effect. Colored dyes were placed around the charges so that the dust from the blast would look extra pretty for the television cameras. And when one of the masterminds of the whole affair executed the unthinkable, grinning fiendishly as she plunged the handle downward, government agents and law-enforcement officials tailed her to a gathering of coconspirators, where cops and bureaucrats alike promptly shook her hand.

Far from being the target of a spectacular campaign of ecoterrorism, the more recent destruction of the hundred-foot-high Marmot Dam on the Sandy River was the outcome of careful democratic consensus-building and sober economic analysis. The dam's owner, Portland General Electric, had to decide if the cost of modifying it to allow endangered salmon and steelhead to pass would pay off against revenues from the electricity the century-old structure generated. They calculated salmon would be worth more in the long run and sent their

CEO to ceremoniously blow up what had become a liability in the face of the growing value of free-flowing water and salmon runs. The company's choice is one that's growing more common every year.

America has fallen into the grip of a dam-removal movement that could well have Edward Abbey twitching in his desolate grave. Unlike the wild anarchist protagonists of Abbey's novels, many hands turn the monkey wrench these days. The recent scene on the Sandy River has been repeated across the nation on rivers large and small. Since 1999, more than 430 dams have gone the way of Sunbeam and Marmot. In 2008, sixty-four dams in fourteen states were slated for deconstruction, most of them smaller projects whose locales provide a cursory tour of some of the nation's more obscure place names: Mike Horse Dam in Montana. Steele's Mill on Hitchcock Creek in North Carolina. Rex Creek Tannery Dam on the Lamprey. The Manner's Run Dam. And if you live on either of the country's coasts, it's doubtful you'll be able to celebrate, much less locate, the dismantling of the Four Hill Flowage project on Big Weigor Creek.

But the dam demolitionists keep setting their sights higher. On the Olympic Peninsula in Washington, work has begun on engineering the removal of the 105-foot-high Elwha Dam and, just upstream from it, the 210-foot-high Glines Canyon Dam, the only obstacles between some pristine habitat protected by a national park and what was historically a healthy run of very big chinook salmon. The paperwork is done on a deal that would remove four dams on the upper Klamath River in Oregon, restoring three hundred miles of salmon and trout waters. Backers are touting the plan as the largest river restoration project in the world.

Unlike many other high profile environmental concerns—say, keeping the chainsaws out of national forests, the ATVs off certain rimrock trails, or oil derricks from taking over a stretch of coastline—the disappearance of the neighborhood dam has been greeted with a palpable sense of overall satisfaction. A post hoc analysis done by the nonprofit American Rivers revisited twenty-six different dam removal projects to gauge their efficacy on economic, environmental, and social fronts. All but one surpassed expectations. Data culled from the nascent science of post-dam stream ecology and geomorphology provides some explanation. Working only with gravity and the available geography, many rivers seem to possess an uncanny ability to settle quickly and comfortably back into a channel, even after decades of diversion or impoundment. After Marmot Dam came out on the Sandy River, which cuts through layers of volcanic silt and carries significant annual loads of glacial flour, it was expected to take several years for the channel downstream of the dam to calm down and reestablish a path. Instead, the work was done over one winter and spring.

If the water knows where to go, accordingly, so does the whole panoply of creatures that thrive on healthier river environs. Benthic macroinvertebrates, those bugs that form the base of the food chain in freshwater, respond quickly to the return of the habitat in which they evolved. Fish in rivers formerly blocked by dams have in many cases rebounded far ahead of predicted schedules and, in some instances, exceeded even the most optimistic numerical predictions.

Dam-removal advocates point to the economic benefits of destroying dams as well. Future maintenance and repair costs are forever wiped from balance sheets. Obviated as well are the needs for expensive upgrades to accommodate newer safety and environmental standards. Free-flowing recreation and tourist dollars supplant revenue lost to navigation or electricity generation. Commercial and sport-fishing economies wrecked by dam construction have been restored. Such arguments, based on data, statistics, and logical evidence, are the basis for a common sense rationale laid out in public forums held over the fate of a given dam. But as sound as the numbers may be, they fail to capture completely the considerable zeitgeist dam removal has come to represent.

In *Encounters with the Archdruid*, author John McPhee describes an early iteration of this larger phenomenon. In contemplating his conversations with pioneering conservationist David Brower, whose most bitter regret was his acquiescence to Glen Canyon Dam on the Colorado in exchange for a promise of no more dams in or near the Grand Canyon, McPhee reckoned that dams come across to people of Brower's ilk as

> disproportionately and metaphysically sinister. The outermost circle of the Devil's world seems to be a moat filled mainly with DDT. Next to it is a moat of burning gasoline. Within that is a ring of pinheads each covered with a million people—and so on past phalanxed bulldozers and bicuspid chain saws into the absolute epicenter of Hell on earth, where stands a dam.

This mythical Sierra Club version of perdition resonates, guessed McPhee, "because rivers are the ultimate metaphors of existence, and dams destroy rivers. Humiliating nature, a dam is evil-placed and solid."

As the full extent of the havoc of a century of frenzied dam building became apparent, it was also becoming clear that the damages were not merely metaphorical.

The planet's cash economies have doled out two trillion dollars building dams in the past century. Since the completion of Hoover Dam, 45,000 more such drain plugs five stories tall and higher have been completed. In *Deep Water*, journalist Jacques Leslie reports that so much planetary heft has been redistributed by dam building that some geophysicists contend the speed of the earth's rotation has been slightly altered and, along with it, the tilt of its axis and the shape of its gravitational field. Sixty percent of the world's two hundred major river basins are dammed, and the water behind them has drowned a landmass bigger than California. The mineral content of the world's oceans has decreased since dams now trap the free flow of river-borne sediments into the sea. The deficit of salty deposits to the oceans has become a deadly surplus on land. Some irrigated farmlands are slowly being poisoned as water diverted to crops evaporates and makes soil too alkaline to grow much of anything useful to humans.

The costs to ecosystems are mirrored by the appalling human costs of dam-building regimes. Since the end of World War II, forty million to eighty million people have been rendered homeless, landless, and jobless by dam construction. The critical mass of those victimized by dams brought the building of new ones to a grinding halt by the late 1990s. The World Bank pulled out of financing a couple of big projects, embarrassed by the fraudulent accounting that showed a positive cost-benefit analysis. To address these concerns, in November 2000, the World Commission on Dams (WCD) published a much-anticipated report. Its conclusion: Large dams showed a "marked tendency" toward schedule delays and significant cost overruns. Irrigation dams rarely, if ever, recovered costs. Environmental impacts were "more negative than positive" and "have led to [an] irreversible loss of species and ecosystems." Large dams' social impacts "have led to the impoverishment and suffering of millions, giving rise to growing opposition to dams by affected communities worldwide." Since the environmental and social costs of large dams had never been adequately measured, the "true profitability" of large dam schemes "remains elusive." The WCD report didn't get too much attention in the States, which seems odd, given the reality of the nation's waterways.

The United States has successfully spread its dam-building gospel first and foremost at home. The U.S. Army Corps' official inventory of dams in the United States lists more than 75,000 projects, far more than the industrious Chinese have managed, one for each day since George Washington crossed the frozen but free-flowing Delaware in 1776. Most of the problems described in the WCD report have an analog and most likely a precedent in America, particularly in the West, where the costs of federally financed and built dams were underestimated, the benefits exaggerated, and the ecologic, social, and cultural havoc thoroughly minimized or ignored altogether. The world's modern dam troubles were America's dam troubles first.

The bad news, as the WCD report makes clear, comes down to this: like clear-cutting, strip-mining, factory trawling, or off-shore drilling, dams are as efficient a modern device as any for unscrupulously transferring wealth from one part of a territory to another, generally to the benefit of the rich and the detriment of the poor. The difficulty with too many dams, the WCD report suggested, is the very problem America was allegedly invented to inveigh against: the potential for tyranny in its manifold forms.

It shouldn't come as too much of a surprise, then, that some of the heirs of George Washington were calling for explosives and trainloads of earth-moving equipment in the States even before Nelson Mandela helped launch the WCD report in the late 1990s. Just one generation after *The Monkey Wrench Gang* made blowing up a dam a couch environmentalist's wet dream, ecological fantasy is well on its way to becoming a reality. The evils of the industrial age, dam removal reveals, may not be so solidly placed as they were reckoned to be just a short while ago. An act once widely perceived to be a crime against property, an offense against the state, has quickly become another practical, effective, and increasingly popular means toward the creation of a better world.

Thus emboldened, the modern anti-dam contingent in the Pacific Northwest has made a modest proposal. Blowing up Sunbeam was a crime ultimately made moot by the construction of eight gargantuan federal dams between the Salmon River and the sea. The Columbia River Basin, of which the Snake and Salmon rivers are a part, was once the most prolific salmon-producing watershed in the world. It has been transformed over the past century into one of the most extensive hydroelectric systems in the world. Surely, a compromise of sorts between these natural and industrial superlatives could be reached. Take out half those dams, the proposal goes—four on the lower Snake River in Eastern Washington—and call it even.

To understand what these latter-day advocates of the River of No Return are hoping to have returned, a clear picture of what was being defended with dynamite that morning in 1934 will be necessary. To accomplish this requires travel north, where it was decided a long time ago that prolific annual returns of salmon were worth more than just about anything else that could be accomplished with water.

2

WHAT THEY'RE SMOKING
IN ALASKA THIS SUMMER

The annual trip to Alaska, in particular, renews a cultural memory of abundance, as fishermen yearly experience exactly what abundance looks like when they encounter runs in the millions of fish and pristine habitat. Their understanding and expectation that such abundance is still possible and desirable on the Columbia remains alive, in part because of the annual renewal of this memory.

—IRENE MARTIN, "Resilience in Lower Columbia River
Salmon Communities"

Nose to the cold window inside a passenger jet out of Portland, I watched what looked like tidal waves of snow-clad mountains roll up to meet the plane at half, then two-thirds, then three-quarters of its altitude. With the rare luck of good weather, dropping from this considerable height into Anchorage is nothing short of sublime. The gorgeous view had taken everyone with a window seat hostage, with faces pressed into every Plexiglas portal. By turning my own attention to the frigid, inhospitable world a few inches outside the thin aluminum skin that protected us, I was also was doing my level best to drown out the boozy prattle of a fellow passenger seated to my left, a portly gentleman I'll call Juneau Bob. J. B. had apparently logged some alcoholic hours in the airport lounge in Portland before getting on the plane and had cracked his second five-dollar Budweiser when he struck up a conversation— or more accurately, began a long soliloquy—on his fishing ex-

ploits throughout the Western hemisphere. Before delving into
J.B.'s shortcomings as a potential travel companion, I should
compliment him on his status as one of those burly salt-of-the-
earth types, a part-time Alaska resident who apparently could
fix or build anything with his own meaty hands, which made the
twelve-ounce aluminum can he palmed look like a six-ounce
juice glass. J.B. was certainly friendly enough, too, suggesting
several choice locales where we might get further acquainted
over rod and reel or, barring that, in one of the strip clubs
where he recommended I kill time between fishing jaunts. But
I quickly decided I would not like to ply any waters with J.B. in
Alaska or anyplace else. The booze seemed to fuel his singular
focus on fishing; all the talk about fishing seemed to make him
thirstier still. He continued to quack about his piscatory prow-
ess as he flagged down a flight attendant for Bud number three.

Besides the incessant talking and drinking, I did not care
much for J.B.'s politics either. He was a staunch, militant, angry white conservative sporting a salacious Sarah Palin fetish
and a deep suspicion of any bit of fact, opinion, or speculation that didn't neatly align with his own worldview. When
he asked me what I was doing in Alaska, I dodged the question by asking him why our Alaska-based airline insists on
painting on the tail of its planes the blurry image of a smirking, stoned Johnny Cash. When he ignored my joke and
pressed me further, I lied to him. "Salmon fishing," I said.
"Salmon fishing's no good anymore where I come from, so
I thought I should try it up here."

"Where you from?"

"Oregon," I replied.

"What ruined it for you?"

"Sea lions," I fibbed once more.

"Your problem isn't sea lions, friend," J.B. retorted. "It's
the government and all their goddamn dams. And you want to
know something else?" I shook my head to mean I did not require any further elucidation from him, but he either didn't notice or didn't care. "The reason you'll find no big dams like you
have down there," J.B. pontificated, uttering "down there" as

if he were making some euphemistic reference to some fungal scourge of his nether regions, "is that back then, the working man in Alaska did not need to go crying to the government for a handout every year like he does today."

In hindsight, the lies Bob and I shared each contain a kernel of truth. I was there partially out of a sense that at least some of his crackpot ideas about government could be validated in assessing the damage done to rivers, but also to participate in what he might judge to be another socialist legitimization of a languid welfare state. On a summer lark, I was headed to Alaska, where citizens have the right to harvest and smoke as much of a certain pleasure-inducing Pacific Northwestern staple as they can get their hands on. I'd been invited to help some friends put up their allotment and quite possibly would burn one down myself in the process. I'm referring not to the infamous strains of cannabis grown in the Matanuska Valley of the forty-ninth state but to the millions of pounds of salmon netted in rivers, soaked in brine, and smoked in shacks from the Kuskokwim to the Kenai and beyond. I was there for the salmon fishing, but not the kind that J.B. had in mind.

Alaska's subsistence economy—those things given, bartered, or otherwise consumed outside the modern, jittery, turbo-powered cash economy—might well be the richest in the world if the unit of measurement were calories. According to the Alaska Department of Fish and Game, some forty-five million pounds of wild foods are processed in Alaska every year. Sixty percent of this annual harvest comes from fish, the vast majority of which is salmon, enough to supply every rural Alaskan with an ample store of sea-based sustenance to coax their bodies through the long dark winter. What's more, the majority of these salmon are caught using nothing more than a net attached to a long stick. Beyond the robust commercial salmon fishery and the lucrative sport fishing industry, there's still salmon left over to feed any legal resident with the wherewithal to stand long enough in the water to get one. This is not only the way it used to be on my home river but on many salmon rivers down the West Coast. I am here, like many fishermen, tourists, trappers, miners, and explorers before me, to witness a semblance of biological conditions that allowed human life to flourish in the Pacific Northwest, a territory the journalist Timothy Egan once defined as "any place a salmon can swim to."

Contrary to impressions from afar, subsistence fishing is not a form of welfare for Eskimos, fur-clad bumpkins, or hard-up homesteaders, nor is interest in it dwindling. The number of participants has grown commensurately with the population over the years. In an age when one can apparently purchase everything at the click of a mouse, what might the attraction be to even a vestige of this way of life? I pose this question to my Alaskan friends, who settled here as much to take advantage of the state's inexhaustible recreational opportunities as to take jobs as teachers. John is an avid whitewater kayaker, whose filmed exploits on the steep creeks draining the Talkeetna and Chugach ranges are harrowing enough to qualify as extreme sport porn. He and his wife, Trish, no slouch herself on skis or paddle, live near the base of Hatcher Pass in the Talkeetnas. A good salmon stream, the Little Susitna, runs through their backyard, though neither one of them cares to fish there, nor have they fished a lick in the Lower 48, despite a stint in the much ballyhooed trout heaven of Montana.

After landing in Anchorage, and mercifully clearing earshot of J.B., I found myself cruising along with John and Trish in a steady July rain, a dipnet and pole strapped to the top of their Subaru. We were headed to the Kasilof River, just south of the mouth of the more famous Kenai. The weather report had not been encouraging. The short opportunity for summer to make an appearance was disappearing fast. "Termination dust," what Alaskans call the first visible snow on the highest peaks that traditionally marks the end of the brief warm season, had fallen on the mountains the night before. To pass the time in the gloomy drizzle, I posed a question. Why spend the precious days of a fickle summer season fishing, especially if you're ambivalent about the process?

"When we first moved here, some people from our church

asked if we wanted to go dipnetting," John recollected. "A lot of people we've met up here do it. Hard to pass up free salmon, I guess. Neither one of us ever goes fishing up here, except for this time of year, when we dipnet."

I pointed out that the notion of free food as the reward for any hunting or fishing endeavor often winds up as wishful thinking when the cost of gas, guns, rods, bait, bullets, flies, and lures is factored in. "This works out pretty well for us," said John, whose cost-benefit analysis of dipnetting includes a few days' lost opportunity to kayak. "We've only been doing this a few years, but the first year we got our limit pretty quickly. Then there's a day of filleting and smoking. Still, it's fifteen fish per head of household, plus ten for every dependent. That's only twenty-five fish, but we wind up with at least seventy pounds of salmon, probably more than that, even after smoking it."

The main quarry for dipnetters is sockeye salmon, whose return to Alaska rivers despite tremendous pressure from commercial fishing is astonishing. This portion of Alaska, the Upper Cook Inlet, produced a run of around 5 million sockeye on its half-dozen major rivers in 2007. The Kasilof, a small river, ushers more than a million sockeye annually into its fecund reaches. This is but a fraction of the more than 40 million sockeye that returned to Bristol Bay. Commercial boats in Alaska brought in 46.3 million sockeye in 2007. The total commercial harvest numbers that year for all five species of Pacific salmon, according to Alaska Fish and Game, adds up to nearly 213 million fish. That year's catch yielded 941 million pounds of salmon, enough to put about three pounds of salmon on the plate of every American diner.[1] Though there is a considerable variety of opinions over whether such numbers constitute a sustainable harvest, Alaska's commercial wild salmon fishery is the largest in the world to be certified as sustainably managed by the Marine Stewardship Council, whose seal of approval is rigorous and relies on third-party verification.

~ ~ ~

Just in time to ply the incoming tide, we arrived on a sandy beach marking the Kasilof River's entrance into the Cook Inlet. Miraculously, the clouds lifted, and that most precious of commodities to Alaskans, sunlight, illuminated the shimmering seascape. Far across the inlet rising in the distance like twin polar mirages, two ten-thousand foot volcanic peaks, Illiamna to the south and Redoubt to the north, watched over the calm ocean.

On either side of the narrow mouth of the Kasilof, forty or so dipnetters lined the banks, each at his or her station, somewhere from knee- to neck-deep in the drink, manning staffs ten to twenty feet in length, at the far terminus of which could be found a hoop or square enclosed with gillnetting. Attending to each of those in the water was a shorebound crew of one to ten people, most often family, whose jobs included clubbing, cleaning, and untangling fish from the net. Many children were involved, learning to gut fish on occasion but mostly learning that requisite skill required of any competent fisher: patience. The fishers themselves looked like neoprene- and rubber-clad pawns in the confused choreography of some wacky landscape art stunt.

The success rate over the past few days looked to have been pretty good. The beach was littered with fish heads, the spoils of caught and cleaned reds. I heard a woman scream, "All right, number twenty!" A man, apparently her husband, waddled toward her with the gait of a baseball manager visiting a struggling pitcher on the mound. He unceremoniously clubbed number twenty, then returned to his chair, cozying up in the chilly sunlight to the cooler and a paperback.

Trish had contracted fish fever. "I can't wait to get in the water. I swear you can feel them around your legs when a bunch of them start swimming in," she told me, donning a yellow dry suit, more commonly used in rafting or kayaking Alaska's arctic rivers. Hers was by far the brightest getup of any other dipnetter in sight. Looking like a walking sunrise, she quickly found a spot in the queue. The latent English major nerd in me wanted to share that Rilke wrote that birds fly through us, not around us; but the platoon of seagulls that were snacking on the ubiquitous fish heads at my feet intimated I should keep this bit of wisdom to myself. The gulls were insatiable, insanely noisy, drunk on fish guts and blood, rioting like victorious warriors as they goose-stepped down the beach in search of more sockeye viscera to devour. Their cacophonous attraction to the banquet created in the wake of successful fishermen was clear enough. But what about the people? In addition to my friends' nonchalance about fishing, I heard the same lukewarm sentiments expressed already by several others shivering in the tide. "We don't like eating salmon all that much," one middle-aged Asian man in a dark green rain suit told me, "but we like coming out here every year."

What kind of fish drives someone who doesn't especially care to fish to do it anyway? Fishing happens for fun as well as profit, but what called me there was its instantiation as a profoundly cultural event. People were gathered along the banks in good spirits for the same reason they gathered along countless similar spots along a thousand other good rivers. To put up food against the coming winter, teach kids where their meals ultimately come from, catch up with the neighbors, strike up a conversation with a stranger, have a drink, court a lover, admire the sunset, affirm a faith in things unseen, or marvel at the mystery of the sea and its riverine veins that connect it to land. In this land of plenty, people also attend to salmon fishing for the same reasons others go to church or the town council meeting or the opera: while there are aspects of these endeavors that can be enjoyable, such events are also seen as part of a larger set of obligations that make the world a better place. The added attraction, of course, is that one of the biological drivers of human culture, appetite, is satisfied in the process. Even a cursory tour of the natural history and life cycle of salmon reveals their adoption by the people of the Pacific Rim as the centerpiece of cultural and religious life as an altogether appropriate choice.

The range of Pacific salmon encompasses half an ocean and the edges of two continents, from Northwest Mexico up the coast to Alaska and across the Bering Strait, then south along the Eastern Pacific Rim as far as Kyushu Island in Southern Japan. Around the Pacific Rim, the seven Pacific species of salmon[2] migrate as far as a thousand miles inland, as high as seven thousand feet, to the high alpine headwaters of their natal streams. Salmon find their long way home from the sea, find a mate, have sex, and die shortly after. On an unpoetic level, this progression appeals perhaps for its obvious contrast to the mammalian way of species propagation, which complicates even the mere prospect of sex by requiring an epic commitment of time and energy for nurturing and raising progeny. Salmon delay their one and only shot at true love until they're swimming at death's door.

Science provides the basis for an understanding of salmon reproduction at once more realistic and inspiring. Wild salmon are exquisitely adapted to their home rivers, which are identifiable in their individual genes. "Fin printing," the sampling of salmon DNA by taking tissue from the rays of the tailfin, maps not only their genetic makeup but their way home from the ocean. The DNA of a North Umpqua River chinook is different from that of a chinook born in the Klamath Basin. This ability to adapt quickly to specific environments is but one advantage of salmon's reproductive strategy. A female salmon, or hen, will lay four to five thousand eggs in the process of spawning. Fertilized by the male, or buck, each tiny egg becomes a kind of genetic trial balloon with the extremely high odds of getting popped and withering into the food chain before reaching adulthood, making certain that none but the very fittest survive. Salmon thus evolve faster than most other creatures. Cognizant of this fact, genetic researchers have come to prefer working with salmon DNA over that of other organisms since alterations to their nimble double helix over successive generations occurs at a faster clip than in most others. The survival strategy of having nimble genes has some drawbacks, too. To maintain this rich base of genetic diversity, salmon need lots of habitat in many rivers, which they don't have everywhere anymore.

Yet there remain a few good places where salmon return to spawn every year. If it's a spot people can get to with relative ease, the prospective mates are bound to have a human audience. The salmon mating ritual seems to possess a deep and mysterious power all its own. Old women giggle, stoic men weep, noisy children fall silent, and certified public accountants confess religious awe at the sight of an active redd, which is the technical term for a salmon nest. Redds are dug by female salmon during spawning season and appear as if someone managed to pressure wash a patch about the size of a picnic blanket in the rocky river bed. In order to dig this bed, the hen will find her home riffle, less than stone's throw from where she was born a few years earlier. She moves up to half a cubic yard of gravel with her tail, rotating her body so that its broad lateral surface works to burrow into the coarse river rock. The eggs she lays must be protected from predators, anchored securely enough that rising water won't blow them out of the gravel, yet close enough to the current to be bathed in oxygen found in flowing water.

Meanwhile, a male salmon, or buck, jealously guards over the redd his hen digs. There may be multiple bucks present in the initial stages of mating, and they are viciously territorial. I've seen two bucks cause such a splashing ruckus they appeared more like fighting ducks. Hens, too, will defend the territory they've chosen for a redd from other hens with a vigor equal to the male's.

To the victor go the spoils: the winning buck then makes coy advances toward the hen, who seems to react with something approaching shyness. There is a palpable, familiar, yet paradoxically inscrutable form of affection between buck and hen that takes place during what is both their opening and final performance. To see it this way only confirms our often tenuous, inadequate but vital connection to the animal world. We might like to watch salmon spawn as much for a sense of the familiar as for contact with a mysterious and unknowable difference. The animals must know something valuable that we have yet to learn, a secret we would someday like to hear.

Those doomed salmon, at any rate, affectionately nudge, rub, and slide their bodies against each other. This all-too-brief courtship seems to arouse the pair; the hen lays her many-thousand eggs, the buck follows, clouding the redd with an ejaculate called milt that fertilizes the eggs. The hen covers the redd with gravel. The spent salmon then part ways, drift into quiet eddies, usually finding a peaceful and agreeable end. Whatever message is conveyed in these moments between mating fish is lost to us. It is in the void of this loss that the fullest implications of the spawning act seems to dawn on human observers: an affirmation of home, the sacrifice to the promise of future lives, the sensuality of both sex and death, and the portions of the ocean and riverine depth known by the fish and unknown to humans. In the enigmatic way of the prophets, through this silent, poignant disappearing act, salmon seem to testify to the great power and fecundity of the marriage of land and water on earth. Plus they taste good.

We eat scads of them long before they ever get the chance to make us swoon to the significance of their long journey home. We eat salmon by the millions, and the miracle of their manifold and myriad returns, year after year, in river after river, in spite of this trespass against them, makes them all the more miraculous, all the more a sacramental meal that people rightfully feel nourishes heart and soul as well as body.

Dead salmon nourish the kingdom's inhabitants just as generously when they meet their natural end in a river as at the end of a fork. At least 137 plant and animal species have salmon to thank for their presence in Pacific ecosystems. The menagerie of flora and fauna ranges from the tiny Trowbridge's shrew to the cumbersome killer whale. Decomposing salmon bodies provide the coniferous forests of the Pacific Rim with a baseline of soil-enriching nutrients, nitrogen in particular, that add girth to the region's spruce, cedar, Douglas fir, hemlock, and pine trees. This boost in turn puts more large woody debris in streams, shades and cools the water, and provides a habitat for the bugs that salmonids eat, which ultimately invites more salmon up these same rivers.

Dead salmon also feed a host of saproprobes, including their own offspring. The eggs laid by a hen will incubate in her redd for two to four months and will be partially fed, along with a host of other macroinvertebrates, by the rotting flesh of their parents. Eventually, the eggs, the color of an alpenglow evening and about the size of a coffee bean, will hatch, the egg sac still attached. They are now called an alevin, appearing as a tiny orange bubble with a fish stuck through it. For the vast majority of eggs and alevin, their brief existence serves as a snack for a host of predators: birds, other fish, and waterborne scavengers of every kind help themselves to the helpless fledgling salmon. The lucky ones grow to a tiny sliver of silver called a fry, though many of them are eaten by another round of relentless predation. Fewer still make it to the next stage of progress, a more familiar trout-looking thing called a parr. Less than 10 percent of a redd full of salmon eggs become parr; these fish are then flushed out to sea on the high flows of spring melt-out. *Flushed* seems the proper term: from creeks near the Continental Divide to the saltchuck in the estuary, a journey of a thousand miles in the pre-dam era took only a fortnight. Under these ideal conditions, young salmon swim near the surface of their rivers, facing upstream back toward their home waters like a wide-eyed kid in the way back of the family station wagon reviewing the road just passed.

In the estuaries they become smolts,* where their metamorphosis from fresh to saltwater creatures is one of the more dramatic in the annals of marine biology. They get bigger here, as one might expect, but in order to meet the demands of the ocean and its saltwater, other dramatic physiological changes occur. For one, the function of the kidneys must adjust. Also, since oxygen concentrations in saltwater are less than in freshwater, a smolt begins to produce a different kind of hemoglobin.

---

* Dr. Carl Schreck, a salmon biologist at Oregon State University, discourages the use of this term. "Salmon smoltify," he corrected me. "When you went through puberty you were not a pube."

Tiny membranes in the gills, salt pumps that keep plasma electrolytes from becoming diluted in fresh water, now reverse their operation to keep these electrolytes from becoming too concentrated. The adult color scheme—silvery sides, dark dorsal surface, white belly—occurs in this stage as well, all the better to camouflage in the depths of the Pacific. Timid creatures in their rivers and estuaries, smolts have to learn a kind of judicious aggressiveness, fishing for zooplankton, shrimp, and smaller fish, while avoiding larger-jawed predators. Many will not learn well or fast enough. Parr and adult salmon numbers are again greatly reduced by predation. Only a lucky few make their way into the ocean and to the adult stage, living stealthy and strong, as long as they are not caught in fishermen's nets, whacked by whale, seal, or sea lion, or otherwise eliminated through an untimely oceanic end.

Then some biological alarm sounds, and, miraculously, these fish begin the trip home. It's quite a swim. In 2003, a steelhead from Idaho's Clearwater River was caught off the Kuril Islands near Japan. Though this journey was exceptional, a loop of a few thousand miles for any Pacific salmon is fairly commonplace. The full story behind the uncanny navigational abilities of salmon has not been uncovered. But there are important clues. Salmon have olfactory senses that make the average canine seem like an agoraphobe with bad allergies, and, similar to migratory birds, may possess the ability to detect the slightest changes in electromagnetic fields.

However they do it, a few adults, hopefully at least two, find their way back to spawn within a stone's throw of the redd where they were born and begin the cycle anew. Such far-reaching and mysterious transformation and travel, and such unfathomably precise navigation, have inspired a healthy, imaginative mythology among many of the indigenous human inhabitants of the Pacific Rim. The Ainu, the Gilyak, the Chukchi, the Koryak, the Tlingit, the Haida, the S'Klallam, the Quinault, the Yakama, and the Yurok, to name but a few, place salmon at the center of their religious and cultural practices. From first-salmon ceremonies practiced by these indigenous people, honoring the first catch of the year, to maritime yarns of the Pacific fishing fleet, salmon have sparked a body of art, literature, myth, rite, and ceremony as varied, if not as voluminous as any of the world's religions. As of late, however, in a more ominous and disturbing prophetic vein, the once-innumerable central characters of this faith are threatened, in many places fast disappearing. Even in Alaska, fears of fewer and smaller fish are coming to fruition.

~ ~ ~

Trish shrieked and cackled with the gusto of a witch who's won a door prize. She'd netted her household's first sockeye of the year. It was John's job to retrieve the unlucky red, dispatch it with a club, and clean it in the surf. "I'll catch and eat them, but someone else has to kill and clean them," Trish explained before hurrying back out to her favored spot. John examined his wife's catch with mild alarm. "These are so small," he said. "I was looking around at other people's fish, and everything seems small this year. Look at this thing; it's barely two feet long."

"I would say it's not even two feet long," I judged.

"They're all like that this year here," said the husband of the couple who'd caught twenty. "It's the cold summer."

~ ~ ~

At a half-dozen different fish camps I visited, the guarded talk turned to the rapid, obvious, and dramatic effects of climate change in the far north. Talking global warming with Alaskans is like talking to a prudish fundamentalist family about their short-haired aunt and her lifelong woman "friend," with whom she lives a few miles out of town. Everyone knows the couple well enough, but nobody wants to talk about them for very long. To dwell on the subject would be to shake a foundational sense of reality loose from its moorings. It's been estimated that climate change in northern latitudes is occurring two to three times faster than in temperate regions. One fisherman told me he'd met a crew of social psychologists who'd booked gigs in Alaska to study how local people were responding to dramatic changes in ecological conditions that had persisted in relative stability for thousands of years. Of course, there's no need to go all the way to Alaska to study something like that.

For those attuned to such things, to visit an otherwise healthy river bereft of its formerly abundant runs of salmon is to confront an inconsolable sense of loss. A proper eulogy for rivers in the West from which species of salmon have been extirpated would run to epic lengths. The extremely abridged edition might start something like this: The Ventura River north of Los Angeles once harbored a healthy steelhead run. The San Joaquin and King's rivers, draining the high peaks of the southern Sierra Nevada, were chinook rivers. King salmon could be seen leaping Salmon Falls in the mountains of northeastern Nevada. Dipnetters gathered to catch salmon at Shoshone Falls on the middle Snake. The Elwha River in the Olympic Mountains west of Seattle held a run of kings that averaged more than fifty pounds, with individual fish occasionally topping a hundred. Grand Coulee Dam cut off a thousand miles of the upper Columbia from some of its largest, wildest salmon, some of which spawned far into Canada.

On the brighter side, the power of salmon to inspire resiliency in humans with whom they share a homing instinct to local streams would be difficult to underestimate. River folk have long gone to extraordinary lengths to honor the connection between sea and soil that salmon embody, and have stalked, fought, and occasionally defeated salmon wastrels of various kinds—dam builders, mine operators, and yokel poachers—with a zeal reserved for hunting down a huckster who'd absconded with the family fortune. In many instances, this analogy isn't far off the mark: in addition to their sacred cultural roles, salmon are worth a lot of money. Alaska's healthy salmon economy is the state's number one employer, worth $1.5 billion annually.

Alaska came to keep more salmon in their rivers than did the bereaved fishermen of the Lower 48, the conventional wisdom goes, for three basic reasons: Fewer people. More habitat. Less development. It's an assessment that's unlikely to provoke serious disagreement. Yet a significant portion of the plenitude of the north owes much as well to some wise, conservative decisions of the recent past. In Alaska, conservation of salmon stocks (stock being the imprecise term for salmon native to a particular stream) is required in the state's constitution. Many of Alaska's sixteen thousand salmon-bearing streams are monitored with annual benchmarks for returning fish as the determining factor in setting catch quotas. Local biologists have the authority to open or close fisheries based on daily progress toward these goals. (Much to the chagrin of our shore-bound dipnetters, seagoing set-netters working fish headed for the Kasilof were being allowed out for a twelve-hour session, approval for which had been given the afternoon we arrived.) This kind of immediate, delegated authority helped salmon harvests rebound from their dismally low numbers in the 1950s, when Alaskan fishing was in sad shape. The year Alaska was granted statehood, 1959, the catch had plummeted to about 20 percent of its current yield.

In addition to letting sound biological practices dictate catch quotas, Alaskans outlawed the practice of open-sea fish farming in 1990. Prior to this, Alaska enacted the Anadromous Fish Conservation Act, which severely restricts logging, mining, and road-building activities in or near salmon habitat. Most emphatically, Alaska has refused dams and the heavy reliance on hatcheries that comes with them. As the Alaska Department of Fish and Game proudly proclaims: "Alaska has been willing to forgo the economic benefits from activities such as hydropower development in order to sustain salmon runs for future generations. For example, although the operation of large-scale hydropower facilities on both the Susitna and the Yukon River were closely examined, neither was built. The wild salmon resource from these drainages was a major reason that Alaska chose the no-dam option." Removing dams may be a good thing, but as Alaska has demonstrated, better to not to have built them at all. J.B., I have to admit, was right.

~ ~ ~

The silhouette of dipnetters beneath the Redoubt Volcano near midnight looked like a gorgeous scroll painting out of Oriental antiquity somehow rendered in Technicolor. The fishers and even the boats headed out to sea were dwarfed by the scale of mountains and water, bathed in the gold and red light of an Alaska midsummer's eve. Trish had fished herself into a mild hypothermia-induced delirium and passed out cold in the car without getting out of the day-glo dry suit. John and I sat up too late and talked more. He's gone native: he and his wife had bought a balloon-tired, four-wheel drive contraption called a Rhino with their oil royalty checks from the state, all the better to access the put-ins to steep creeks. My friend confessed to me in the dusky light that he might like to kill a moose someday. When last I had seen him, his marriage to the shivering fisher asleep in the car was on hold and suffering from the gravest of doubts. Trish had taken a job in Alaska and moved without John. Many good friends counseled him not to follow her, to cut his losses, and resign himself to a painful though practical split. John is a sensible man without religious conviction; nonetheless, he heard some sage old whisper of a voice tell him that he at least needed to try to patch things up with his prospective ex. Alas, their divorce did not work out.

He recalled that dipnetting was one of the first things he and Trish did together after his arrival: "We caught our limit in a couple of hours. And we had to work together, with some help from friends, cutting fish, making the brine, staying up all night smoking. We had no idea what we were doing. But at the end of the process, we had all this food to share." It was certainly not the mere act of killing, smoking, and eating seventy pounds of sockeye alone that rescued their wrecked relationship, nor could anyone of sound mind recommend fishing as a balm for even the mildest forms of marital trouble. Both had an inkling that things might work out before they borrowed their first dipnet and drove to the Kasilof. But they agreed the experience affirmed a fragile hope.

Still, perched on a chilly beach in the land of the midnight sun, I couldn't help but wonder what inspires these kinds of miraculous little resurrections. I thought of the way that parents care for children; writers edit one another's work; barflies keep one another comfortably buzzed; fishermen, hunters, and farmers keep one another stocked with milk, vegetables, and meat; or bucolic villagers raise a barn. Religion, ceremony, and ritual have certainly formalized this impulse, yet its ultimate source remains something of a mystery, such a contrast to the abjectly cruel ways in which we also manage to treat one another. Could it be the planet's generally agreeable climate, its edible grains, its fleshy quadrupeds, its fecund seas and soils that oblige us to continue on with the general scheme of promoting life in spite of the certain annihilation of the self?

For whatever reasons, we are drawn to the great concentrations of bird, fish, and mammal life, a healthy portion of which has always been found in and near the river. Some sensibility that a good meal might be derived from such a spectacle is undeniably a part of the attraction. But another chord deep in the human psyche resonates with the earth and its fellow creatures in the presence of the beautiful. This might be why people who don't care about fishing come to net a few salmon. They may well be obeying some imperative, be it biological, ethical, or aesthetic, to eat beauty, to make it part of their own blood. "Ingesting the sense of the world," wrote Robert Bringhurst, "of which we are made, and to which we return, is just as essential to life as digesting its physical substance."

Somewhat bewildered by the fact that I've flown all the way only to be scrawling in my notebook far after midnight, John arose from his chair. "Sleep," he called out, like a thirsty man crying for water. He paused to drink in the scenery one more time. "Good to see you friend. The river life," he pronounced before unzipping the tent door, "is a good life."

Trish got up before any of us the next day. For an hour or so, the fishing was so good it could have just been called catching. At one point, she netted three sockeye in one dip. The line of fishers shouted and laughed and struggled to control their nets as they jumped to avoid swamping their chest waders in the wake of fishing boats headed to the docks. A lone man in a T-shirt and faded red baseball cap shivered in the oncoming tide; a father and son just laughed and moaned and slapped the water with their net when they lost one.

A thin, stooped, weather-beaten old woman in tattered waders stopped to talk to us, her big nose dripping from the chill of standing in the surf. The rain had returned. "I'm eighty years old," she announced to us without a prompt. "I just want one fish tonight. I'll be happy with just one."

I kept thinking of her on the cloud-enshrouded flight south, back to the land of lowered expectations.

3

FEED WILLY

Tell me what you eat, and I will tell you what you are.

—ANTHELME BRILLAT-SAVARIN

As it turns out, humans aren't the only creatures that have developed a ritual appreciation for a specific kind of fish. Among their many vital biological roles, salmon seem to be the preferred object in certain rituals of cetacean sitophilia. In the coastal waters of the Pacific Northwest, food play prefigures foreplay for young male orcas, who will occasionally show off their fresh chinook catch by donning the silver carcass on their heads like a clownish hat, a novel party prank apparently performed for the usual reason, to garner the attention of a female. The gag has been observed as part of a rare but impressive gathering of orcas called a superpod, where the whales circle up like inductees to some waterborne contra dance, some of them "spyhopping," a move that puts them vertically in the water, heads poking above the surface. Like all good festivals, the atmosphere appears to be sexually charged and may facilitate pairing with a desirable mate. But for orcas who've evolved to spend summers in and around Puget Sound, the future isn't so bright. The superpod and its attendant salmon-hatted clowns have been declared in danger of a final performance in some gloomy summer of the near future.

The orcas in question make their summertime home in the waters of the sound, alias the Salish Sea, upon whose eastern shores lie the giant cities of Vancouver and Seattle, as well as the middling metropolitan regions of Bellingham, Olympia, and

Tacoma. It's a peculiarly Pacific Northwestern mix of greasy industrial and urban plush on the one hand and mossy, tangled, wet, conifer-shadowed, salt-tinged, and wild on the other. Like people, the orcas seem to like it here, especially in summer, though the area is undergoing rapid change, much of which does not bode well for either species.

The specific group of orcas that makes its home from May through October in Puget Sound, known as the "southern residents" (as opposed to the "northern residents" whose range begins in the waters off the northern half of Vancouver Island and extends up to southeastern Alaska), joined Snake River salmon on the endangered species list at the beginning of 2006. In recent years, as chinook salmon returns have plummeted, the group of orcas that spend their summers cruising Puget Sound are showing clear signs of stress. Soon after they were awarded endangered species status, three whales from this group, known to their human watchers as Jellyroll, Raven and Hugo, dependable visitors to these waters throughout their lifetimes, were observed to be suffering from malnourishment-induced duress. Raven left, abandoning her four-month-old calf to a likely death. She was seen with a condition known as peanut head, where the area behind the blowhole that normally stores fat becomes hollowed out and sunken. Hugo had ribs showing, and Jellyroll parted ways with her two-year-old. These are signs of calamity, especially since orca mothers and their offspring usually stay together for life.

Somewhat more dramatically than famous potatoes, Puget Sound orcas used to be nurtured by Idaho waters. Historically, as a group of whale biologists pointed out in a letter to federal fisheries managers and a federal court, the Snake River drainage provided the lion's share of salmon upon which this special breed of killer whales relies. The scientists explained it well in their letter:

> As research scientists who have spent decades studying fish-eating . . . killer whale (Orcinus orca) populations in the Pacific

Northwest, we are writing to call for your leadership to protect and restore abundant, self-sustaining populations of wild salmon and steelhead in the Snake and Columbia Rivers. . . . As you are well aware, Columbia and Snake River salmon are intricately linked to the continued existence of Puget Sound's Southern Resident killer whale population. . . . The Columbia and Snake River Basin was once the world's most productive salmon watershed. Today, only about one percent of the historic number of these fish return. Yet the Columbia-Snake Basin still holds more acres of wild river than any watershed in the lower 48 states. It is this opportunity for salmon and steelhead recovery that we must take advantage of as the last best hope for a substantial increase in prey availability for Southern Resident killer whales during the critical winter months. . . . We call on your agency to follow the science and include the removal of the four lower Snake River dams as an essential element of a real recovery plan for these extremely important fish stocks.

Whale researchers were calling for the removal of dams seven hundred miles from Puget Sound, connecting the dots between the health of riverine and ocean ecosystems, and outlining a strategy for saving two endangered species with one brave management choice.

Ken Balcomb has been studying the sound's orcas for four decades. Since 1976, Balcomb has lived on the windy west side of San Juan Island in Puget Sound, where his living quarters occupy the top two floors and his office and lab the bottom level of a home overlooking the Haro Strait, just north of the Strait of Juan de Fuca, the watery neck that connects Puget Sound to the Pacific Ocean. With a salt-and-pepper beard and a compact physique well-suited for the task of maintaining balance on the deck of a boat, he looks like Captain Ahab, had the fictional commander of the Pequod done a research stint at the Scripps Institute of Oceanography. But Balcomb, a laconic fellow, doesn't talk like Ahab or a scientist caught up in the vernacular of his trade. He's blue-collar plainspoken, the guy you might wind up playing in a friendly game of pool or the neighbor who could help you replumb a kitchen sink. This unassuming nature belies his considerable experience. Over three decades, Balcomb has archived thousands of detailed pictures of killer whales. The photographs allow researchers to learn the identity of individual whales by the pattern of their saddle patch (a swirl of lighter-colored skin near the base of the dorsal fin on their black backs) and track changes from year to year.

From these records, for instance, Balcomb knew that the oldest southern resident, until her death in 2008, was Lummi, a female born around the time of World War I, making her the grande dame of known orcas in the Pacific. Balcomb has offered his expertise to the international whaling commission and has been outspoken against the U.S. Navy's plan to bathe the world's oceans in whale-eardrum-splitting sonar. And he will also tell you pointedly that the whales upon which he has built his career are in trouble.

"We're seeing more of them dispersed," Balcomb told me on an uncharacteristically calm, clear summer day after picking me up at the dock in Friday Harbor on San Juan Island. "Orca cultures," he told me from behind the wheel of a well-worn white Ford Escort station wagon, "are about hunting. For the southern residents, it's about hunting chinook salmon. The hunting is getting tougher every year. They're swimming longer and longer distances for less and less food. A lot of their food, especially in winter months, used to come from Columbia and Snake River salmon. And the fish they've relied on, the chinook, is in an even deeper fix."

Balcomb believes the only way to rebuild populations of chinook is to take drastic measures in addition to tearing out the Snake River dams: "I think a ten-year moratorium on the commercial catch, and even severely restricting sport fishing, is probably the only way to save these stocks." The sentiment is not likely to sit well with the fishing industry or even some conservation groups, whose public relations campaigns often feature commercial fishermen in a united front with restaurant owners and chefs promoting wild salmon dinners as consumer-savvy and ecologically virtuous. But to Balcomb and other cetacean scientists, the ethics and aesthetics of food culture are not the exclusive domain of humans.

The ability of orcas to teach and learn distinct languages is part of their culture. The southern resident killer whales were the first animals in the brief legal history of the Endangered Species Act to make the list partially on the basis of this distinction. The phenomenon of cultural transmission in the animal world has been well-established. A 2001 paper published in the journal *Behavioral and Brain Sciences* put it this way: "The complex and stable vocal and behavioral cultures of sympatric groups of killer whales appear to have no parallel outside humans and represent an independent evolution of cultural facilities." The southern residents are one of those sympatric groups (that is, a genetically distinct group within their species, having evolved so without the influence of geographic isolation). The designation, I related to Balcomb, seems potentially problematic. Certain animals are to be loved because they are "like us," as smart as a scientist, family-oriented as the Mormons, red-blooded as a farm boy, trainable as a retriever, sympatric as a convention of gregarious Irish tenors, and, *voila*, cultured as the French. Balcomb shrugged. After studying orcas for most of his life, it's neither surprising nor controversial to him that their intelligence ranks them among higher-order primates. He's more interested in uncovering as much as possible about the migrational, dietary, and linguistic particulars of the orca's relationship to the chinook and less about what could be properly labeled as "culture." Each orca pod, a matrilineal family group of orcas, uses a characteristic dialect of calls to communicate, with certain calls used in common between pods. The calls used by these whales are unlike the calls made by any other community of orcas and can travel ten miles or more underwater. (One southern resident pod's call sounds for all the world like a cat meowing.)

Curiously, although the southern residents have been

known to eat rockfish, flatfish, chum, pink salmon, and even the occasional seal, they generally turn up their enormous black snouts at every other culinary possibility but chinook when this fattest and largest species of the genus *oncorhychus* is available. Research by Canadian whale biologist John Ford found that where chinook are present, they comprise over 90 percent of the southern resident's diet. It takes two hundred to three hundred pounds of seafood a day to keep an adult orca well-fed, about what you might expect for a creature averaging twenty-five feet in length and weighing in around five tons. Thus, the eighty-six members of the southern resident orca pods consume in the neighborhood of eight hundred thousand king salmon a year, a nice round number the Columbia River once easily provided with millions to spare. Not so anymore: in the seventy years since the first dam was built on the Columbia, no annual return of kings has exceeded a million, and the average over the decades is far short of that.

"If you plot the graphs of chinook populations and southern resident numbers, they track each other," Balcomb told me. "We've established a low-threshold number for a population of Pacific salmon that will feed these whales. Every time since '76 that their number dips below that threshold, the number of orcas drops with it. This last time it happened is what finally got the whales listed. It may be, that if these orcas are to survive that they would have to teach themselves to eat from a different food base."

Despite an uncanny mammalian intelligence, learning to eat differently isn't a task these orcas are ready to take on quite yet. In three decades worth of studying the southern residents' diet, no evidence has yet been garnered verifying even an hors d'oeuvre of Fraser River sockeye salmon, though the sockeye outnumber chinook hundreds of times over in the summer waters of Puget Sound. Acquiring a better picture of just how profoundly chinook influence the southern residents has taken some scientists to some extraordinary lengths. Researchers in the field have spent considerable time cruising Puget Sound in the halcyon days of summer, soaking up the occasional sunny days, contemplating the dual majesty of the Cascade and Olympic ranges, plying the idyllic blue waters—for orca shit. When a killer whale drops a deuce in the Salish Sea, the specimen eventually floats. Other forms of digestive whale detritus—fish scales, barfed-up bones, the spoor of prey species—present an incomplete sketch of a standard orca diet, along the same lines perhaps as inventorying the wrappers and paper cups in a car to determine the health of a human diet. So whale researchers sport a long-handled skimmer on board their research crafts, putting years of expensive scientific training to use by netting the proverbial Baby Ruth bar in the swimming pool, though this inept metaphor quickly falls to pieces when one considers the diffuse consistency of the quarry, known as a "fecal cloud."

On the way to Balcomb's windswept abode in the white wagon, there were some appointed rounds to make, one of which was the suburban analog to the collection of whale waste. We made a deposit at San Juan Island's landfill and recycling center, where we witnessed the processing of the more voluminous waste stream of the human species. Maybe we were prompted by the sight and stench of the island's refuse, but for whatever reason, as we recycled what we could and threw away what we had to, the discussion turned to the toxic threats facing the world's marine mammals, the southern resident orcas in particular. It was not an upbeat conversation and eventually came back to a kind of double jeopardy that the lack of salmon creates for Puget Sound's orcas.

As with many other species of marine mammal, the orcas suffer from a modern-day malady plaguing many creatures that precariously occupy the narrow spot atop a food pyramid. A toxic stew of heavy metals, pesticides, industrial chemicals, and even human pharmaceuticals[1] work their way up the food chain, beginning with the tiniest invertebrate organisms and ending with the largest red-blooded mammals. The poisons increase their concentrations at an exponential rate for every step up the trophic staircase. This insidious process, called biomagnification, has bestowed polar bears, seals, whales, sharks, dolphins, porpoises, and some humans with contaminants stored in their respective fat layers at such concentrations that their continued survival might one day soon double as the premise for some low-budget science fiction melodrama.

On a January day in 2002, a female orca washed up on the shore of Dungeness Spit on the Olympic Peninsula. Levels of PCBs in her tissues were so high that the device used to measure such things had to be recalibrated to get an accurate reading. A male accompanying her, possibly her son, was so distraught by the death of his beloved that he insisted on repeated efforts to beach himself next to the deceased. For three days under the media spotlight, volunteers gave aid and comfort to the despondent male, finally coaxing him back into the ocean. The female, it turned out, had the highest level of PCB contamination ever found in a killer whale.

The tragedy exposed a grim spiral of starvation and toxicity. The extra energy orcas spend searching for sparse numbers of chinook works against them beyond ordinary stress and fatigue. Burning their stores of blubber dumps those fat-sequestered toxins directly into their bloodstreams, decreasing resistance to disease as well as their odds of successfully calving and rearing their young. Restoring chinook runs would keep them from burning poisonous fat reserves that exacerbate the deprivations of hunger.

The high-profile incident also drew much-needed attention to the polluted waters of Puget Sound. It appears likely that some form of a multibillion-dollar, decades-long restoration effort might mitigate some of the damage done. But any effort to recover the sound, Balcomb pointed out, might well be cancelled out by the sheer crush of people resettling in the region. By 2015, somewhere between five million and seven million new residents will settle in the Interstate 5 Corridor between Vancouver, British Columbia, and San Francisco. Sustaining this rate of growth over a century—some 10 to 12 percent annually—translates to more than double the number of people sharing what is currently orca and salmon territory. By 2100, if the current pace holds, some 145 million people will call some portion of this area home. "If all these people want a big house, two cars, and a green lawn," observed Balcomb, "the salmon along with the rest of us will be in fairly deep shit. And we haven't even talked about climate change."

I asked Balcomb if the explosion of wealth in the Seattle area—there are now more millionaires in the metropolitan area than there are school teachers—has benefited funding or awareness of the plight of the sound's killer whales. "There's more money," he replied. "And there's more people out looking for and at whales. But people's attention, really, is elsewhere. There are more distractions than ever. I doubt that five percent could tell you there are resident killer whales in the sound, and far less than that would know anything about what they eat."

Given these stark political realities—not enough people outraged or paying attention as a host of postmodern maladies work to snip the threads of the region's intricately woven bounty, a tapestry of mountains and trees, salmon and whales, river and ocean—why call first for the dams to come down? "I'm not against cleaning up the sound. But even if it were pristine tomorrow, the southern residents would still be short salmon," explained Balcomb. "In historical times, the rivers of the West—the Sacramento, the San Joaquin, the Klamath—all contributed to this huge biomass of chinook in the coastal waters. But the Columbia was the big horse of all these rivers. I think most fish biologists, if they were to give you an honest answer, would say that those dams have to come out for those stocks of chinook to have a chance to recover. These fish are about a hundredth of their historic abundance. The only other hope is for some benevolent politician to come along and declare all remaining stocks of Pacific salmon to be reserved as whale food. And that's not going to happen. But we're going to have to do something. Probably something drastic. And soon." Time, as Balcomb and other scientists are warning, is running out. But the science also hints there's room for recovery.

~ ~ ~

In the past few decades, an assortment of historians, ethnologists, anthropologists, and biologists have been engaged in an emerging effort to recollect how particular landscapes appeared before hordes of European settlers arrived. Such reconstructions are not an exact science. Seldom do single organisms, much less entire ecosystems, behave in predictable ways. Partly because of this conjectural tendency, most hypotheses and conclusions are hotly contested. But these kinds of endeavors have quite practical applications. As fledgling efforts to restore some ecosystems to a semblance of their former glory are undertaken, accurate estimates of former and current carrying capacity need to be made, in part to work up an educated guess on whether or not there's enough resilience left in the system to justify restorative measures. On the Columbia and Snake, researchers are trying to determine the accuracy of Balcomb's estimate that chinook salmon are at a hundredth of their former plenitude.

It's generally been agreed that the Columbia was once one of the world's most prolific salmon rivers. But pegging this greatness to a number is another matter. In 1986, an advisory panel created to help restore salmon estimated the pre-dam, pre-European invasion size of the Columbia Basin salmon run at ten–sixteen million strong. This figure was simply the middle of estimates they'd sought, but it quickly took on a life of its own, and has been oft-repeated by both sides in the salmon wars for twenty-five years. Recent evidence points to the possibility that the number was higher—perhaps much higher. In a roundabout way, Balcomb's decades-long, detailed study of the culinary habits of orcas, along with other voracious salmon-devouring predators, provides some startling evidence for an annual count of salmon far beyond what's been previously considered.

According to Bill McMillan, a salmon scholar whose career spans roughly the same period as Balcomb's, a good place to begin deciphering a more precise historic salmon count would

be with the current acrimonious battle against another voracious chinook-devouring mammal. Sea lions make their way
north in the spring, foraging for salmon along the coast. Voracious salmon fishers themselves, somewhere between a few doz-
en to almost a hundred follow the runs up the Columbia, living
high on the hog at the terminus of navigable water for them, the
base of Bonneville Dam. Like orcas, sea lions display a keen intelligence. Also a favorite of circus-style aquarium trainers because of their brains and adaptability, California sea lions have
been trained by the U.S. Navy for underwater mine detection
and equipment retrieval. Their cousins, the Steller sea lions,
or northern sea lions, while not yet booking any gigs with the
military, are similarly adept hunters, and annually a few find
their way to the base of Bonneville Dam. The burly pinnipeds
run afoul of the authorities and fishermen for their propensity
to devour large quantities of what fishermen proclaim without
any sense of irony is an endangered species. The U.S. Fish and
Wildlife Service conducts a costly hazing and relocation of the
offending beasts, a program that went awry in May 2008 when
six sea lions died in traps at the base of the dam.

Media portrayals of this problem give the impression that
the sea lions swimming so far upriver is a recent development,
something along the lines of feral cats in cities or hospital strains
of bacteria resistant to disinfectant. But the historical record revisited by McMillan and some other scientists indicates several
species of seals and sea lions always have swum in the Columbia. Moreover, they used to swim even further upriver in much
greater numbers than seen in recent decades.

In the fall of 1805, and again in the spring of 1806, Lewis
and Clark were continually impressed by the abundance of
"seals," "sea-calf," and "sea otters" they saw on the lower river,
remarking in their journals at the routine sight of them at several locales on the lower Columbia. Their astonishment was
echoed by explorer David Thompson in 1811 and naturalist
John Kirk Townshend in 1836.

"For Lewis and Clark to be impressed by the abundance of

an animal, it would have meant animals in the thousands," Mc-Millan wrote in a letter to the editor published in the *Columbia Basin Bulletin*. "This is remembering they had seen herds of bison, the remarkable abundance of salmon, and waterfowl so numerous near their Vancouver encampment they complained they could not sleep at night for all the noise. . . . I have estimated from consistently similar descriptions of 'numerous' and 'great numbers' that seals and sea lions in mixed presence cumulatively numbered 100 per river mile, with concentrations of thousands of them together at those points where salmon and steelhead were particularly concentrated."

The dent this historic population level of seals and sea lions must have put in salmon runs yields some enlightening numbers. Celilo Falls, fifty miles upstream from Bonneville Dam, in the pre-dam era divided the lower river from the middle and was then as far as sea lions could travel, two hundred miles from the ocean. If McMillan's estimate is in the ballpark, then up to twenty thousand seals and sea lions swarmed the lower river in its prime, consuming five–fifteen million salmon every year.

Sea lions eat their fair share of fish, but no creature has eaten more salmon than humans. McMillan's investigation of the record of this catch also provides evidence for abundance that defies conventional wisdom. As a general rule accepted by anthropologists and historians, indigenous catches accounted for 30 to 50 percent of a given salmon run's total size. Even if the Indians along the Snake and Columbia rivers took half the salmon, descriptions of tribal salmon harvest offer the possibility of abundance beyond the currently accepted range. What follows is a sample of the historical record McMillan has amassed, accentuated in a few instances by other sources:

On October 1, 1882, forty thousand to fifty thousand salmon were counted by a local settler, drying on racks at the confluence of the Little Spokane and Spokane rivers. October was long after the June–August peak of spring chinook and steelhead returns, and this was but one of several major fishing sites on the Spokane system. This river was never surveyed for salmon productivity before the first dams were built in the early 1900s. Based on records of direct observation, this midsized tributary of the Upper Columbia may have produced a million fish per year.

Downstream of Spokane, on the upper main-stem Columbia at Kettle Falls, Indians from at least a dozen tribes from as far as British Columbia, Montana, and Idaho gathered to distribute more than a million pounds of salmon caught there annually. A daily catch of nine hundred to seventeen hundred salmon, March through October, was hauled in three immense thirty-foot wide baskets that hung below each of the most frequently used salmon passage points. Fish that missed the mark with their leaps fell into the baskets. The numbers caught in the nets did not include the many salmon and steelhead that were speared each day by the individual fishermen. One observer estimated that by this latter method, a thousand adults present at the falls each received three salmon each day during the peak of the run from June through August.

Fishing with nets and spears at falls and other natural constrictions, Indians of the Columbia Basin caught and consumed millions of pounds of fish taken at hundreds of good fishing places. On far reaches of the southern end of the drainage, Shoshone-Bannock people fished at Shoshone Falls on the Upper Snake River. Archeological exploration has uncovered evidence of a native fishery at Dagger Falls on the Middle Fork of the Salmon River, a thousand miles from the ocean. Nez Perce Indians gaffed "June hog" chinook salmon—weighing some sixty to eighty pounds—on the Imnaha River in Eastern Oregon. Spring chinook were taken in this way at Salmon Creek Falls in northeastern Nevada.

The most cosmopolitan of any of these sites was Celilo Falls, on the lower Columbia near what is now The Dalles, Oregon. Archeologists have estimated that Celilo has been continually inhabited and fished for as many as twelve thousand years. Historian Katrine Barber, in her book *Death of Celilo Falls*, echoes the sentiments of many other historians, anthropologists, and Indians themselves when she declares Celilo, until its inundation by

The Dalles Dam in 1957, one of "Native America's wealthiest trading communities." People from all over the Western hemisphere gathered to trade, gamble, race on foot and horseback, marry, celebrate various rituals and customs, and, of course, fish. How many fish did they catch? In October 1805, William Clark counted "107 stacks of dried pounded fish in different places on those rocks which must have contained 10.000 ^{lb} of neet fish." As Clark noted, this was after the salmon run was over for the season; the store of fish was interspersed with stowed nets and lines used in dipnetting. The previous day, Clark had studied the methods by which salmon were preserved:

> I observe great numbers of Stacks of pounded Salmon neetly preserved in the following manner, i. e. after [being] suffi-[c]iently Dried it is pounded between two Stones fine, and put into a speces of basket neetly made of grass and rushes better than two feet long and one foot Diamiter, which basket is lined with the Skin of Salmon Stretched and dried for the purpose, in this it is pressed down as hard as is possible, when full they Secure the open part with the fish Skins across which they fasten th[r]o. the loops of the basket that part very securely, and then on a Dry Situation they Set those baskets the corded part up, their common custom is to Set 7 as close as they can Stand and 5 on top of them, and secure them with mats which is raped around them and made fast with cords and covered also with mats, those 12 baskets of from 90 to 100$^{lbs.}$ each form a Stack. thus preserved those fish may be kept Sound and sweet Several years, as those people inform me, Great quantities as they inform us are sold to the whites people who visit the mouth of this river as well as to the nativs below.

In his elegiac but slyly comic autobiography, *When the River Ran Wild!*, George Aguilar Sr., a Warm Springs Indian now in his late seventies, provides some insight into how much food William Clark observed cached on the banks of the Columbia that year: "A 100-pound basket of *itk'alik* [fifty 20-pound salmon]—mixed with dried berries, four gallons of dried roots, some eels, and dried salmon heads—could easily feed a family of four for four months of the winter." Salmon dried this way, as Clark noted, was not only cached for the winter but leveraged as a valuable trading commodity. It was also widely considered to be delectable. According to Aguilar, the tribal version of the parlay between Lewis and these Chinook Indians differs from the oft-repeated version, which says Lewis and Clark ate dog because the taste of salmon was offensive to them.

"The salmon has high energy value," writes Aguilar, "with protein, carbohydrate, calcium, and thiamin. So valuable was the salmon in the spring of 1806 that the Lewis and Clark party could not persuade some of the River Indians to sell them some. They had to settle for eating dogs and horsemeat."

Offering alternative renditions of historical assumptions about the gustatory preferences of the Corps of Discovery, Aguilar also paints a vivid picture of the bustling fishing season scene around Celilo Falls. At times, the fish were so thick in the river that fishermen could *hear* certain runs of salmon as they splashed their way upriver.

Once past Celilo, salmon returned in large numbers to every creek in the watershed where decent habitat was available. For a few decades, white settlers were able to witness the bounty of one of the world's greatest salmon rivers. Creeks you could hop across held astonishing numbers of salmon. In 1915, a run of thirty thousand spring chinook made its way up the Tucannon River in Eastern Washington, one of the smaller tributaries of the Snake. The terms *salmon fishing* and *Nevada* are not widely associated today, but the salmon were there throughout the northern portion of the state in significant numbers as recently as the 1930s.

As late as 1924, along the middle Snake near Weiser, Idaho, an enterprising young man ran a fish wheel in the summer months and sold 2,400 salmon to restaurants in the Boise area. In the 1870s, a similarly ambitious entrepreneur caught 75,000 sockeye salmon in the Idaho lakes that used to harbor them, selling the fish to miners working the mountains nearby. He described the number of sockeye as running into the millions.

In the four or five miles downstream of Windermere Lake, near the Canadian headwaters of the Columbia, there were so many large chinook salmon redds in the late 1800s that they blocked steamboat travel up the river to the lake as the flows receded in late summer. One English businessman in the Upper Columbia headwaters nearby indicated there were "millions" of salmon carcasses lining the riverbank when he first came to the Columbia in the 1860s. The Hudson's Bay Company, in the same part of Canada, recorded the sale of hundreds of bearskins that reeked of the salmon on which they'd fed.

In 1883, almost 43 million pounds of spring chinook were brought to the docks in Astoria, Oregon. This number merits further scrutiny. In those years, nearly all of the salmon commercially caught went to canneries, which bought only spring chinook because of its high oil content. Fall chinook, and every other species of salmon, went upriver unharmed and uneaten. If twenty to thirty pounds works as an operational average for these early season chinook during these years, the take—not the return but simply the fish caught—was between 1.4 million and 2.1 million kings. The fall run easily doubled this number; escapement (the number of fish that make it past all the hooks, nets, and claws, beaks, and teeth of predators) of springers probably equaled it again. The catch that year represents in some ways an ecological debacle—over the long haul, the fishery could not sustain this high percentage of take—but nonetheless illustrates the plentitude that once existed.

Pending a final draft from McMillan on salmon's former abundance, historians and biologists can speculate freely on where his final estimate might wind up. McMillan has a ballpark figure but warns against laying the blame for the demise of salmon in the Columbia solely on the dams. "A lot of the upper basin runs seemed to be headed for extinction long before the dams went in," McMillan wrote in an e-mail. "Around

1883 [the year of the record spring chinook harvest], there was a huge collapse, most likely due to overfishing." Nonetheless, in spite of multiple threats to the continued existence of salmon, McMillan, like Balcomb, believes that dams, and especially fish hatcheries that proliferated as a means of mitigating losses due to dam construction, are the single biggest obstacle to recovery of the species.

The ultimate aim of the study, McMillan noted, is not some nostalgic meditation on loss and regret. The point is to identify stocks that are still potentially recoverable. The upper basin runs, especially in the smaller tributaries of the Snake River, have been written off or ignored entirely, contends McMillan, because carrying capacity has been grossly underestimated. In the meantime, that ballpark figure of his is worth considering: annual totals for all five species of salmon in the Columbia-Snake system may have been as high as 35–50 million.

Whatever the precise number, the Columbia was, as Balcomb put it, the "big horse" of Western salmon rivers, the leading contributor to the "biomass" of chinook upon which orcas and a host of other creatures fed. As we zipped around Friday Harbor, from hardware store to post office to gas station, it was clear that Balcomb's concern was less with the past than with the present and future. "One percent is generous," he acknowledged. "You don't even have to dig into the past to see that. The southern residents are really a remnant population."

As we made our way out of town toward his home and office on the other side of the island, he continued: "Their numbers were cut at least in half. Even though there are many strains of chinook listed as endangered all along the coast, their numbers continue to slide. Most of these fish, even in a decent year are hatchery fish. Hatcheries are run to benefit fishermen, not whales or any other creature. That's at least one good argument for bringing back the integrity of the river system. If you're interested in restoring the function of the whole ecosystem, there has to be enough wild fish out in the ocean to feed everyone, the whales and seals. And the fishermen, too.

"But when you look at what's happening on the Columbia and the collapse of salmon populations everywhere south of there, I think the ocean fishermen could accept a decade off. They're already doing it anyway now, more or less. At least then, their kids and grandkids might have a chance to fish. Even with all that, we might not get the salmon back anyway, and the southern residents might well go with them."

We pulled into the long driveway of Balcomb's home over-looking Haro Strait. I asked what has been the most memorable thing he's ever seen in the panorama out his big kitchen window, part of a large open room that includes a grand piano tucked in among some impressive whalebones. "I saw the Navy testing their big sonar system right out there when they knew there were whales nearby," he told me. "I used to see fishing boats. Saw them take a lot of salmon right offshore. One boat would set a net, pull it, then another would come. You don't see that anymore."

He paused before shifting the talk back to the future. "The quiet consensus of some scientists on salmon is that, by 2100, they're extinct," said Balcomb. "Best-case scenario is a few token runs of hatchery fish. The causes are agricultural development, dam construction, timber harvest, urban development." It's daunting, I confessed to Balcomb, to think that the best possible efforts to mitigate any portion of the ecological mess we've made will simply be triage; that a slim hope for maintaining whatever biological integrity remains might be the best option we have.

I asked him if he ever loses sleep over these seemingly intractable modern problems. "Well, I used to," he said. "I've kind of taken the view a friend of mine has. This guy was heir to a fortune, money out the ass, and felt he could change things by giving to a big conservation group. He put in countless hours in addition to his millions. But at some point he stopped because he could see that things were coming unraveled so quickly that it wasn't going to make much difference. That's the way the world is designed for now. At least, the human side of the world. The biological and ecological side had a design that worked really well for a long, long, long, long time. This little add-on we've created to that is clearly not sustainable. But there's a small group of people who've done really well by it and want to keep things that way. And then there's the mass of people who aren't doing as well, but they feel like they're stuck with it. For the time being there doesn't seem any way out of it, no matter who you vote for."

Balcomb's phone rang. A neighbor was calling to report a flurry of orca sightings off her backyard deck. "She's an advocate for us," explained Balcomb. "Her place is always open to anyone who wants to sit on the deck and watch for whales." We hopped back in the beat-up white wagon and, a few minutes later, were greeted at her place by Balcomb's brother Howard Garrett, who runs the nonprofit Orca Network. Its multifaceted mission involves coordinating whale-watching volunteers, leading whale-watching expeditions, and organizing a campaign to free Lolita, a southern resident orca stolen from Puget Sound in the late 1970s to serve at the pleasure of the aquarium industry. "Hey, bro!" Balcomb greeted his brother with a hug. "I guess I should've known you'd beat me here."

The enthusiasm for whales is a family affair, and with a serendipitous reunion underway, the view from the deck seemed all the more spectacular. To the southwest across the water, the high peaks of the Olympic Range watched over the Strait of Georgia. A slight breeze seemed to usher in a touch of the golden light of late summer afternoons. With no stiff winds, the sun seemed downright hot. I was handed a pair of binoculars and given the rundown on the easiest way to sight a whale. "Just look for any place there's a bunch of boats that congregate in a hurry," explained Howard. I caught sight of a pair of orcas in a matter of minutes. The distance, of course, was far too great to make out any sort of detail that might indicate whether these were southern residents. But the attention from a flotilla nearby made the odds fairly certain that someone would be able to add to the voluminous portfolio of orca photos Balcomb started more than thirty years ago.

We couldn't stay long on the deck. Balcomb still had to

complete his appointed rounds, and I had a ferry to catch back to Anacortes. But seeing the whales seemed to dissipate whatever gloomy pall had been cast over the day by the subject matter of our conversation. Then it occurred to me that I was the only one that felt gloomy. Balcomb, while not happy at all about the plight of rivers, ocean, and orcas, nonetheless stuck me as somehow quite content with the world. As we parted ways, I queried him about the source of his satisfaction. "The bright side," Balcomb offered as a parting shot, "is that biological processes will eventually win out. Whether we choose to change, or whether things continue to slide, we'll see a turn of events that will ultimately be corrective. And if we don't choose, well, there's a lot of resilient organisms out there that will adapt. They'll literally swim past the corpses to spawn or find something to eat. But humans might not be a part of that scenario."

On the ferry back to Anacortes, the timbered slopes of the islands, the azure skies, and the calm blue waters of the sound made it seem preposterous that there might be anything wrong. With whales and salmon still available to pique the curiosity of biologists and thrill tourists, it would have been a tough argument to make on a day like that one that the biological systems that sustain life were coming unraveled, much less that humans might not be around to take in future splendor. It just wasn't the kind of day that allowed much room for dwelling on any possible end for the human experiment. Much more pleasant to drink in the scenery and try to put a finger on the kinds of things that allow us to hope, in spite of the evidence, as one poet said it, that "humanity might outlast civilization."

Back on the mainland, I got stuck in Seattle traffic on I-5 near Boeing Field. I crossed under an overpass marked West Marginal Way. Boeing, I knew, is one of the preferred customers of Columbia River dam-generated electricity. Much of the power to build its planes comes from the same river that connects killer whales to salmon and Idaho to the Salish Sea. Shimmering Puget Sound itself reflects the distant ripples of these intricate connections, how the full cargo belly of a

747 might influence the empty belly of a whale. Dams on the Columbia and Snake have so dramatically altered the annual pulse of river-borne minerals to the sea that surface salinities on coastal waters from Alaska to California have changed. Hydrologists describe a river deprived of its mineral- and-soil–composed sediment load as being "hungry." Like whales, hungry rivers go looking for food, the literal salt of the earth they carry.

The Salish Sea is starved for a taste of the mountains. Its orcas are hungry for a taste of chinook, preferably with that added pinch of salt.

4

BUTTE CREEK

There is not a "fragment" in all nature, for every relative
fragment of one thing is a full harmonious unit in itself.

—JOHN MUIR, *A Thousand Mile Walk to the Gulf*

As researchers like Balcomb or McMillan will confirm, dams kill a lot of salmon. They destroy vital habitat. And once the flood-gates are closed, unfettered access to any remaining free-flowing habitat upstream of the dam is also lost. Warm reservoir slack-water behind dams may be fine for bass, bluegill, crappie, and sucker fish but isn't so great for salmonids. Dams also have a tendency to concentrate pollutants as well as predators in loca-tions that further jeopardize the existence of whatever brave, hardy fish survive in such altered habitats. But in a world cur-rently playing host to a frightening array of unfolding envi-ronmental catastrophes, the conclusion reached by Balcomb's philanthropist friend is worthy of further consideration: with ecosystems rapidly unraveling, how can tearing out a few dams make any damn difference? While this line of inquiry will be dismissed by devout conservationists as a nihilistic cliché, the question holds implications not only for the future of further dam removals but for the larger environmental movement.

Even as America became the first nation in history to pass a body of ambitious environmental legislation in the 1970s, the assumption persisted on the part of government and industry for a long while that ecological standards could always be met out of what was considered to be the surplus of a given natural resource. Minimum flow targets for streams would be achieved *after* irrigators and purveyors of hydroelectric power got what they needed. Wilderness areas could be carved out of high-elevation "rock and ice" habitats, leaving merchantable timber to fall to business as usual. There was always next year's snow-pack or one more tract of forest to clear. But in an era of exponential population increase, and in an economy that's predicated on perpetual growth to furnish profit as well as meet reasonable human needs, it's been too seldom considered that inevitably the surplus will run out. Tearing out dams could become a way to blast a hole through this dead end, figuratively if not literally.

Ecology teaches that a healthy, functioning ecosystem maintains its principle—its salmon, trees, soil, clean air, and water. Ideally, all other activities live on the interest. Coming to terms with this concept, which is at the heart of all the talk of sustainability, has proven to be an insurmountable obstacle for a fragile ecological conscience. A cursory examination of the balance in this metaphoric account book so disturbs a significant portion of its examiners that they're apt to draw the quite understandable conclusion of anyone facing bankruptcy proceedings: fuck it. Trash that spreadsheet and run up the credit card. The repo man's coming anyway.

Part of the political potency of dam removal is its fast-acting effectiveness: restoring free flow to a creek quickly and dramatically restores ecological principal and, in doing so, hope that some kind of amicable settlement with the earth might be reached. Moreover, the work of consensus-building and the restoration work that usually follows a toppled dam is not only good for stream biology but good for democracy. Though none has gone uncontested, dam removals so far have been characterized as much by the art of compromise and cooperation—with the recognition that some meaningful gesture of reconciliation with the biological integrity of a watershed, even if it's small, can produce dramatic results. Give a river an inch, and it might very well return a mile. Such miraculous comebacks are heartening. Whether they might ever overcome the damage still being done is quite another matter. But in the work of recovering ecological debts, a small creek in California has become an intriguing study in the reinvigoration of the most liquid form of local biological capital.

For its miraculous recovery of a salmonid species that probably never should have survived there in the first place, it's hard to beat the story of Butte Creek. Though it originates pure and clean on the southern flanks of Mount Lassen in California, in its lower reaches, near Chico, these waters for half a century have been a testimonial to how not treat a watershed if you want it to bear a healthy load of aquatic life. To begin with, surviving a journey through the Sacramento-San Joaquin River Delta is no mean feat for any aquatic creature. Massive pumps on the south end of the delta can reverse the natural flow of water, putting six million acre-feet of water into southbound aqueducts that slake the thirst of some twenty-five million Californians. (An acre-foot is enough water to cover a hypothetical acre of ground a foot deep.)

This heavily diked, diverted, channelized, and polluted habitat was once one of the richest marine ecosystems in North America. The delta was the crown jewel of the 450-mile-long Central Valley, comprised of the valleys of the Sacramento River to the north and the San Joaquin to the south, draining the southern terminus of the Cascades as well as the wetter west side of the Sierra Nevadas. As opposed to the Mississippi or Nile deltas, where the narrower point of their respective triangle-shaped alluvial fans is situated inland, broad end extending to an estuary and the open sea, this delta empties the Sacramento and San Joaquin rivers into the north end of San Francisco Bay through a relatively narrow notch in the coastal mountains. Thus inverted, the wider inland base of the triangle in spring once became an immense wetland covering up to eight thousand square miles. The entire Central Valley, uniformly flat and just a few feet above sea level, was a giant marsh in spring; upland, it was one big, wet, flower-laden prairie, home to millions of birds, herds of tule elk, and sprawling stands of oak. Its madrigal of tributary creeks would be packed with chinook salmon and steelhead smolts, some of which would return a few years later to spawn in streams as high as seven thousand feet in the Sierras. These days this grand ecosystem has been so decimated that the conservation group American Rivers cited the delta, or what is left of it, as a prime factor in nominating the Sacramento River as the nation's most endangered waterway in 2009. Nature responded to the designation with its usual recalcitrance, adding to the litany of biblical plagues being endured by the state by supplying it with yet another drought.

A few chinook do still survive the perilous passage through the erstwhile estuary into the main-stem Sacramento. About half a million hatchery kings, and around one-tenth that many wild ones, are all that's left of a river system that was second only to the Columbia River in ushering salmon to and from the Pacific, producing five million of them annually. A disproportionate number of those remaining wild fish have recolonized a creek even a sage biologist might overlook. The wildly successful recovery is due in no small part to the removal of four dams on lower Butte Creek.

Near the bucolic burg of Verona some eighty miles up the Sacramento River, close to its confluence with the Feather River, a few brave wild salmon will pause at the outflow of an industrial irrigation and storm-water gutter known as the Sutter Bypass. Somehow distinguishing the scent of their home waters amid a slurry of agricultural pollutants, these fish instinctively enter the canal, a forty-mile-long, rail-straight half-pipe seemingly designed by sadistic engineers to omit any features that salmon might find accommodating. With any luck, they'll pass the west foot of an ancient eroded volcano known as Sutter Buttes that marks the highest point in the Central Valley. Flanking a couple golf courses, they'll pass the Butte Slough Outfall Gates, where the winding Sacramento River bends east to meet—almost—the northwesterly vector of the artificial bypass canal. This spot was actually the original confluence of Butte Creek with the Sacramento. The entire flow of the creek is diverted into the Sutter Bypass channel, the confluence artificially displaced back to Verona, collecting agricultural runoff and rainwater from farms along the way.

Eventually, the confines of the artificial channel give way to the foothills of the Sierras, and Butte Creek is freed to pursue its natural course. Pine, eucalyptus, blue oak, and native grasses cover volcanic soils. The creek bends to obey the path it has carved over centuries through a fifty-million-year-old lava flow. Provided they've survived the swim of a lifetime, Butte Creek salmon are free to enjoy the relative sanctuary of these waters unencumbered by improvements made by enterprising Californians, the fish spawning and rearing their young each year as they have for centuries.

It's within this refuge that I've come to meet Allen Harthorn, director of the Friends of Butte Creek, who has graciously agreed to give me a tour of chinook spawning hereabouts, even though it's started to drizzle. With river sandals loosely strapped onto his feet, wearing shorts and a hastily donned windbreaker, he leads the way from his trim yellow house down to the creek a few hundred yards away. It is raining steadily, the first significant precipitation in months. Having been raised in the heart of salmon country, which comes with a healthy skepticism against all things Californian, I'm poised to catch a glimpse of what I've found on other creeks around the state—anemic low flows, underwear or socks encrusted with baked mud and balled up against a bank, ubiquitous beer cans, heaps of brown-colored foam in the eddies, and perhaps, hanging on for dear life, a four-inch trout nosing around in the deepest, coolest pool available. My first clue about how wrongheaded this idea would wind up to be was the piles of bear scat we stepped over even before Butte Creek came into view. "They must eat a few of these salmon," Harthorn says, "but they're pretty quiet about it. I've been down here for twenty-five years, and I've seen a bear just a few times. Never in the daytime."

Bearanoia leaps from the unconscious of the suburban hiker at the slightest provocation. Just as Harthorn testifies to the rarity of his seldom-seen bruins, I hear a splashing and believe for an instant I might be witnessing the end of my gracious host's quarter-century of bearless days. But the sight is even more fantastic: two female spring chinook in aquamarine-colored water, splashing and battling it out, still in the competitive phase of the mating ritual. I flash back to the palm trees I saw passing through Chico a half hour before. My rain-dampened brain, attuned to its home waters, for a moment cannot equate palm trees with salmon.

Back in 1987, Harthorn remembers, chinook here had just about had it. Only fourteen spawned in Butte Creek that year. In the ensuing years, some fishermen, including Harthorn, got together and convinced Pacific Gas and Electric (PG&E) to release a little water from a dam it owns up on the West Fork of the Feather River via canal to Butte Creek, during the critical summer months. The next year, there were more than a hundred salmon. PG&E agreed to another shot of water from the West Branch into Butte Creek in 1992. Again, the chinook population bumped dramatically in response. The idea that Butte Creek might be something special began to sprout, first in the fishing community. "It was really a commercial fisherman, Nat Bingham,[1] who was getting out into the community around Chico, knocking on doors, telling people, 'C'mon, we want to keep these fish off the endangered list, here's how you do it,'" recalled Harthorn. "Around 1995 was really a galvanizing year for the diverters [farmers] downstream. They could see that with a little work, the stream could really kick out fish. All of a sudden, there were enormous sums of money to improve the creek. CALFED, U.S. Fish and Wildlife, and the Central Valley Project Water Association dumped money wholeheartedly into Butte Creek. Then the West Side Irrigation District came up with money to remove Three Point Dam. And eventually, even the Metropolitan Water District [in Los Angeles] kicked in because they always want to be able to buy West Side's water."

Those four low dams were eventually removed. Along with

screening diversion intakes and providing fish passage at dams that stayed, all told, some $30 million has been spent on Butte Creek restoration. The salmon responded with exponentially increasing gratitude. In 1998, the California Department of Fish and Game estimated more than twenty thousand wild spring chinook returned to Butte Creek. "It was probably more like thirty thousand," Harthorn says. "There were so many fish that I don't think Cal Fish and Game really had a method to count them accurately. It wasn't until carcass counts were done later in the fall that we figured the number was probably low."

The abundance of 1998 turned out to be no fluke. A kinder, gentler passage through the Sutter Bypass channel was apparently the only invitation a few thousand more chinook needed to make it to the cool clear waters of Butte Creek Canyon, which now hosts the largest wild population of salmon in the entire Sacramento River system. In each summer of the past decade, Butte Creek has averaged a return of nearly ten thousand chinook. More wild salmon returned to Harthorn's backyard creek in 2008 than just about anyplace else in California, including fish hatcheries on the Feather, American, and Sacramento rivers. Though much improved, Butte Creek still faces some dire trouble.

Harthorn's tour takes us to a deck he built overlooking a large holding pool just upstream from his house. It's a place that's come to mean a lot to him over the past fifteen years, and he looks after it as if it were a part of his family. We watch as a big male descends into the deepest slot in the river. The concentric rings made by his poking about the surface break up the uniform pattern of raindrops dappling the pool. "Temperature here is still the biggest problem we have," Harthorn says, noting that in most summers the pool into which we're gazing tops 70 degrees, a thermometer reading generally considered to be lethal to salmon. "These fish either have adapted somehow to tolerate higher temperatures or they're finding pockets of cold water. Maybe right down there where that big one just went is one place. When it gets really hot, I think about rigging

up some sort of system to shade this pool, like a big awning."
I survey the square footage Harthorn is talking about shading
and guess it's a third of the area of an infield tarp. "I hope I don't
have to engineer something like that," he sighs, "but the day
may come."

In the meantime, Harthorn procured a grant from the Environmental Protection Agency to explore the relationship between water temperature and salmon survival, funds that have
several graduate students from the University of California at
Davis shuttling in and out of his driveway. "I was trying to help
them get pictures of bears eating salmon carcasses with remote-
sensing cameras. Problem was, every time the flash went off
the bears would attack the camera. So we got some infrared
cameras, and so far this fall no casualties," Harthorn relates. "I
can help with that kind of thing, but my main role here is just to
raise awareness. And I think I can claim I've done more of that
than just about anyone else."

The year 2009 was the first of the last ten that tempered Harthorn's optimism about the viability of the major restoration of
Butte Creek salmon. Fewer than three thousand salmon made it
back. The number is especially depressing to him since there has
been no commercial salmon fishery off the coast of California
since 2006, when, in the face of wholesale collapse of chinook
and coho populations in the Klamath and Sacramento, fishing
boats from Coos Bay, Oregon, to Monterey were idled to help
rebuild the runs. The federal relief package for out-of-work
fishermen along a seven-hundred-mile stretch of the Pacific
coast cleared the $200-million mark but didn't come close to
covering the losses to salmon-dependent coastal communities
and didn't recover any salmon. The California Department of
Fish and Game estimated the cost of the closures in that state
alone came to $270 million.

To Harthorn, this might mean his Butte Creek fish have become increasingly endangered at sea. The lack of salmon may
be so dire in the Pacific, he thinks, that too many of those that
remain are feeding orcas and other predators that depend on

their abundance. "We had a great turn out for our film festival this year," Harthorn says about an annual fundraiser his organization holds, "and I think it was partly because it was at a brewery [Sierra Nevada Brewing in Chico, which supports Harthorn's work] but mostly, I think, because people are concerned we're reaching an endgame with our salmon population. I used to think because these salmon are wild [that is, not raised in a hatchery], they were better equipped to avoid predators in the ocean. I still think that's true, but there may be so few of them left in the ocean that they can't feed their predators there." Harthorn's grave concern for his home creek's salmon, it occurs to me, resonates with Balcomb's worries over starving whales.

Harthorn knew that lethal temperatures for salmonids were more or less a constant threat during summers on Butte Creek. But it wasn't until 2001 that anyone saw how devastating the tepid water could be. The run was coming along swimmingly. Then one of the driest summers on record hit hard. Somewhere between 3,500 and 7,000 chinook died before making it to their spawning grounds. The next summer was even worse. More than 10,000 salmon carcasses littered the creek bed before they had spawned. The immediate problem lay not in pesticide-laden water or the industrialized delta but upstream in the middle of the Butte Creek refuge at the Centerville Head Dam, which reduces flows on the creek in summer by 70 to 80 percent.

Having been granted reprieve from the maze of weirs, low dams, and diversions built for agriculture, the portion of salmon's existence compromised by hydropower began to draw a lot more attention. All the water diverted from the upper reaches of Butte Creek goes to spin an antique turbine in the Centerville Powerhouse, a facility that was state of the art back when it was dedicated in 1899. "It's a very inefficient operation," Harthorn says. "It takes thirty to forty cubic feet per second [of water] to keep the thing spinning. But they have to dump half that water down a spillway to keep the turbine moving efficiently. And for that, all PG&E gets out of it is a quarter to a half a megawatt of electricity. They might be able to use that in the daytime when peak energy use occurs, but no one needs it in the night, and the turbine spins 24/7. For what they're spending here, they could mothball the dam and put in a solar array, which would give them power when they really need it anyway."

PG&E supplies power to fifteen million people in Northern California. Its network of dams and turbines forms one of the largest privately owned hydroelectric systems in the world, connecting sixteen different river basins. The system features more than a hundred reservoirs and sixty-eight powerhouses, producing enough juice to meet the needs of four million homes. Hydro also provides the lion's share of PG&E's renewable energy portfolio, a key asset as the company, along with the State of California, tries to meet ambitious carbon reduction goals.

The company netted $1.34 billion in 2008 but long before that had become a pariah in its home state. On the heels of deregulation in the mid-1990s, PG&E went from being a large, dependable, exclusively Californian utility to a national corporate behemoth, owning thirty generating facilities in twenty-one states, that provided 5 percent of the country with electricity and gas. Deregulation allowed PG&E to finance this venture to an unhealthy degree with ratepayers' money. Californians labored under the delusion that their power and gas bills were covering the cost of upkeep and maintenance of PG&E's power infrastructure. But in April 2001, the company declared bankruptcy—sort of. While robbing ratepayers to finance its expansion, PG&E spun out its most profitable, California-based generating facilities into a subsidiary called the National Energy Group. It made another subsidiary out of its old self, the traditional utility Californians were used to dealing with for their gas and electric services. The generation facilities, under the newfound freedoms granted by deregulation, were no longer subject to government-mandated price caps. National Energy Group was free to charge the same extortionary rates to utilities that other wholesalers like Enron were getting. When one prong of the rapidly expanding corporation, the old PG&E, filed for bankruptcy, it claimed it had racked up $9 billion in debt trying to keep up with wholesale power rates. Half of the debt, in fact, was owed to their sister subsidiary. In a practical sense, the company simply owed itself money. But the debts, as often is the case in the accounting of big corporations, had been neatly walled off from the profits. Some accused PG&E of filing bankruptcy to leverage a huge government bailout of the debt, to prevent its ledgers from bleeding red from the wounds inflicted by chewing off its own leg.

Fed up with these kinds of shell games, some Northern Californians tried to get some other entity to provide heat and light. Public power in California is generally cheaper and sometimes more environmentally friendly than what PG&E can offer. So when some citizens of Yolo County (where lies Davis, just north of Sacramento) maneuvered to transfer their services from PG&E to Sacramento's nonprofit Municipal Utility District, the for-profit provider spent more than $10 million to defeat the measure, which didn't pass.

What might help PG&E regain a modicum of trust, offers Harthorn, is a warm-and-fuzzy public relations campaign with some substance to back the good vibrations. "We've suggested from the beginning that the very best thing PG&E could do would be to shut down the [Centerville] powerhouse and provide full flows to Butte Creek," says Harthorn. "Increased flows produced dramatic jumps in population both times that it occurred first in 1983 and then again in 1992. It would cost less than what they're spending to keep it running. And it would pay dividends in good public relations, far more effectively than any advertising campaign could." The suggestion seems to be anathema to Centerville's owners. PG&E has spent $14 million, along with a share of its credibility, to get the Federal Energy Regulatory Commission (FERC), the agency responsible for issuing licenses to private dams in the United States, to relicense the Centerville project. Most of the good Butte Creek spawning habitat, PG&E is betting, is below the Centerville Project. According to PG&E, even if the water in the Centerville Canal were rededicated to Butte Creek, salmon would simply climb the fishway to get above the dam, take a quick peek around, realize there wasn't much room to reproduce, and then be too dumbstruck or exhausted to move back downstream. The water above Centerville, PG&E also claims, is too warm in summer to do salmon any good.

This theory, notes Harthorn, is not much more than that. There's no convincing data to suggest that this is what would happen, a point the State Water Control Resources Board (SWCRB) took pains to point out to the FERC in a letter that summarized the state's environmental assessment of PG&E's preferred option. The water behind Centerville is warm, Harthorn and others observe, because the shallow reservoir bakes in the summer sun four months out of the year. "They simply made up hypothetical statements," Harthorn emphasizes. "It's all just designed to keep the powerhouse running. It's not designed to benefit the fish. And I think that when we start with a different flow regime, PG&E is going to have a really hard time explaining how more water puts these fish in jeopardy."

That different flow regime, recommended by state and federal agencies, calls on PG&E to provide more water, implementing, on an experimental basis, the conditions the company claims will hurt salmon. But what it also buys the company is time. It will take at least three years, beginning in 2010, to study whether or not more water is good for the wild salmon of Butte Creek, giving credence to a cynical sense that Pacific salmon might well be studied to death before meaningful recovery begins. "In the meantime, there are other tributaries upstream that would support more salmon," Harthorn says, "but for now we're stuck with just this one population in the main stem. And they're dealing with a temperature issue biologists are already saying is critically bad for them. And this is the best wild population in the state of California."

Like the creatures he watches after, Harthorn is patient but determined. He believes the company may yet be convinced to do the right thing. "The problem is," he explains, "there's no manager of theirs, or anyone from their board, who's looking at this from the standpoint of long-term relationships with their customers and shareholders. My hunch is that they only ever hear from their hydromanagers, who've been told 'maintain production at all costs.' Look at what PacifiCorp is doing up on the Klamath.* I would guess that someone looked at the outcome of those dam removals in terms of the company's relationship to the public. It's an easy equation. The dams cost more money than they're worth, so you might as well capitalize on the good press you get from tearing them out. And if you don't, you're basically forcing them to try and adapt to your vision of what is best for these fish, when, in fact, if you just get out of the way, the fish will figure it out."

For Centerville's owners, the "easy equation" is a hard number to read. PG&E, for its part, claims it is not opposed to removing the dam. It only wants its $14 million investment in relicensing to pay off and a chance to see if there might miraculously be any truth to the idea that more water is bad for salmon. Bill Zemke is PG&E's relicensing coordinator on Butte Creek. In a telephone interview, he acknowledged the fact that any self-respecting utility is going to be loathe to give up a practically carbon-free source of power in the years to come, even the relatively puny amount that comes out of Centerville. He acknowledged too that there was no hard data to back the contention that adding the water above Centerville to Butte Creek's flow would hurt salmon. He said he didn't think taking out Centerville was outside the realm of possibility—it just wasn't going to happen anytime soon. I mentioned that Butte Creek now harbors the largest wild run of salmon in the Sacramento River system and reference the breakthrough on the Klamath. Why wouldn't PG&E be interested in a similar deal? PacifiCorp, owners of the Klamath dams, will pay little to noth-

---

* In the fall of 2009, PacifiCorp announced that it would sign on to an agreement in principle to remove four dams on the Klamath River beginning in 2020. Details remain to be hammered out, but if the dams go, the Klamath would become the largest river restoration project in history.

ing of the demolition costs and be relieved of the cost of upkeep for the dams, all of which are newer and much larger than Centerville. "We've been a part of helping bring back salmon to Butte Creek, and we'll continue to be a part," Zemke explained. "But we also think the number of fish in there now is unnaturally high. Six thousand is probably a reasonable number. And most of the good spawning habitat for that many fish is below Centerville. Having said that, I don't think we would rule out taking out Centerville at some time in the future. There are lots of considerations, though. For one, the powerhouse is on the National Register of Historic Places. Would you try to make a little museum out of it? I don't know."

While the company ponders whether the National Historic Preservation Act should trump the recovery of endangered wild salmon, the upshot of PG&E's indecision is maintenance of the status quo while federal and state scientists explore the curious thesis that more water will be bad for salmon. They're also asking the public, including the scientific community, to accept at face value their contention that the fish are "overpopulated" in Butte Creek, despite the fact that salmon once had access to a much greater percentage of the Butte Creek Basin. The move implies as well that the company is willing to pit salmon recovery against climate change to protect their interests. By contrast, decommissioning Centerville would compromise less than 0.25 percent of the company's hydropower capacity and would, if the company displayed even a hint of shrewdness in negotiations, probably be offset with public money to ease its minimal pain.

Why a company whose interests would be well-served by giving up a museum piece like Centerville won't consider the option, at least in the short term, is a subject of relentless speculation on the part of the Friends of Butte Creek and others sympathetic to their cause. The quick, easy diagnosis: the prospect of the onset of perpetual drought. A nervous sense that climate change is settling in over the Sierras for a good long visit permeates every discussion over water everywhere in the state. Anyone with control over water under the circumstances will quickly realize the folly of showing weakness.

Yet beyond the fear of a warming world, PG&E's inflexibility over Centerville Dam and California's water crisis in general have roots longer than any well-adapted desert plant you'd care to name. The heart of this age-old conflict is not really about fish, crops, or power generation. It's about water as a means of control over wealth and, ultimately, over other people.

Hydrology is a complex subject—as complex as the ecology of political and economic power. One of the more insidious arrangements of that power, as any political theorist can tell you (provided you can sit still that long), is that its implementation often dramatically departs from the virtues espoused by the entities that have acquired it. From the vantage point of the present, such contradictions are easy to identify. The Inquisition—what with kangaroo courts and the burning at the stake of the heretics and all—certainly seems a little out of line with what the Catholic Church hopes to accomplish. Legal slavery undermined ideals of justice and equality in a free and democratic society. Burning villagers out of their homes and sprinkling them with napalm as they fled was, in hindsight, an ineffective strategy for winning the hearts and minds of the Vietnamese. But historical lessons clear in retrospect may seem more ambiguous in the moment.

The centuries-old practice of diverting water, ostensibly for the greater good, is no exception to this phenomenon. Whatever it was in good intentions that paved the road to hell also seems to have had a hand in constructing large, centralized irrigation schemes that coincide with some of the more oppressive political regimes in world history. As Don Worster describes in *Rivers of Empire*, this is no coincidence: it is the hallmark of what he calls the "hydraulic society." Mesopotamia, India, and China all built grand water-storage, diversion, and distribution systems, all of which required centralized and often brutal bureaucracies to maintain and operate. Corralling nature, especially in arid places where a fear of scarcity made it easier to manipulate people, the practice of water control became a prerequisite for strong-arming the citizenry. Other historians have as deftly laid out the particulars of water politics in antiquity, but Worster was among the first to so thoroughly apply the concurrence of water control and political hegemony to the present milieu of the American West. The picture he paints is not one many Westerners would care to own. It is a full account of a government operating with a perverse, single-minded dedication to first serve the rich and well-connected at virtually any cost.

An unholy alliance between private capital and federal authority formed on western rivers in the early part of the twentieth century. The scale of power and control it provides over water remains unlike anything the world has seen before it. Two federal agencies, the Bureau of Reclamation and the Corps, radically replumbed virtually every watershed west of the 100th meridian. The potent combination of industrial technology and growing military and economic might made many Americans unquestioning allies of the Bureau of Rec's water-development slogan through the post–World War II years: "Our Rivers: Total Use for Greater Wealth." The "water-hustlers," as Worster describes them,

> would lay their hands on virtually every river and tributary in the region, obliterating entire watersheds in a rage for "comprehensive, multipurpose water development." They would insist, with a sincere, breathless urgency, a frantic, intense will to believe in which was mixed the crassest self-interest and patriotic promotion, that without more and more water, death itself was stalking the land. Their anxious need to get more water, to expand their manipulation of nature, was so intense it became a kind of totalitarian impulse—a drive to capture and hang on to every single drop that fell on the West, allowing nothing to elude their tight control or stand as a challenge to their supremacy.

In tandem, the theme of financial insolvency kept nipping at the heels of the ideal of "Total Use." The Central Valley Project

(CVP), the Bureau of Rec's flagship project, is the nation's largest federal water supply system. Its twenty dams and reservoirs, 1,437 miles of canals, 192 miles of drains, and array of pumping and power generation facilities cost the federal government $3.6 billion to construct. Part of the bargain was that farmers would pay back more than $1 billion of this cost within fifty years of its completion. But as of 2002, the majority of the CVP's patrons were still delinquent debtors. Sixty years after signing up for the water, farmers have repaid only 11 percent of the construction costs. Of the nearly two hundred CVP water districts, 167 have paid less than 20 percent of their construction tab. Twenty-three have paid nothing at all. Unlike other credit criminals, CVP farmers are not being charged interest for late payment. Theirs is not the only obligation that's gone unmet.

Promised as a self-financing venture that would be funded out of repayments from farmers, by 1936 the Bureau of Rec was in such arrears due to lack of farm payments that it took its first appropriation from the Department of the Treasury's general funds, courtesy of Western boosters in Congress. By 1950, its line-item appropriation had jumped to $314 million. "And those sums," writes Worster, "did not begin to indicate the enriched diet on which the agency was now feeding, for also in the thirties it discovered that the generation and sale of hydroelectric power from its western dams was the very elixir of bureaucratic life." Individual wealth and government authority federalized the rivers of the West, making costs public and benefits private.

Federal and state subsidies to the Central Valley's 6,800 farms come to $416 million annually, a gift from federal taxpayers that is bestowed not so much on family farmers but on corporate farm ventures. Measured in terms of total water allocation, the top 5 percent of agricultural water users, or 341 farms, receive 49 percent of the water from the CVP, enough water to supply 2.3 million households for a year. This represents a water welfare check averaging $513,000 per year for each of these enterprises. The thirstiest 10 percent of farms get 67 percent of the CVP's water. The largest, Woolf Enterprises of Fresno County, takes 29,000 acre-feet, an annual gift from the public of $4.2 million.

Like free water, the advent of cheap electricity from dams also came with a promise of a greater good for the many, but this, too, has fallen far short of the intent. If you track this modern problem back to its roots in irrigated agriculture, it's clear that even the language of Western water control was saddled with contradictions from the start. The bureaucrats at the Bureau of Rec and its sister agencies have labored to entwine the public's sense of their dam-building mission with visions of amber waves of grain, a pleasure boat on a glassy mountain lake, the fruited plain, the cost-of-living wage adjustment, a securely lit and electrified castle for the honest working man to call his home. Theoretically, free water and free power were to create a rising tide of economic security that would democratically lift all boats. But in practice, none of what has been "reclaimed" comes without great cost. Rivers and land wantonly destroyed; wildlife and people relegated to a quietly reinforced state of inequality and oppression.

Before leaving the Golden State, I wanted to test Harthorn's contention that salmon know perfectly well what to do when human beings will humble themselves long enough to step aside. Along his stretch of the creek, he'd built more than one place where potential human observers might get out of the way to simply watch. At the beginning of our tour, I'd noticed a makeshift ladder he'd constructed out of scrap lumber, nailed into the trunk of an immense canyon oak, allowing adventuresome tree climbers a stealthy vantage point from which to study chinook. After we said our good-byes, I snuck back down to the creek, climbed the tree, perched on an ample limb, and spent the better part of an hour watching fish getting ready to spawn.

Those same two hens I saw at the beginning of my Butte Creek tour were still fighting over space for a redd. When the smaller one bit the bigger one, she retaliated by pushing her challenger down into the shallow riffle right below my tree limb. I watched her gasping as she lay on her side, the current spinning her slowly down creek. Finding a rare wedge of sanctuary, hemmed in on one side by civilization's insatiable demand for cheap food and on the other by cheap power, the survivor turned and went back to digging her redd.

5

ENERGY VERSUS ETERNAL DELIGHT

Cheapness and ignorance are mutually reinforcing.

—MICHAEL POLLAN,
The Omnivore's Dilemma

When the Snake River dams are finally laid to rest, the credit for giving the idea its first public voice will still belong to a Boise piano teacher named Reed Burkholder. Nothing in Burkholder's house would give you even the faintest clue he devoted a decade of his free time to crafting, then promoting, the detailed economic analysis that would prove to be the shot across the bow in the battle to have these dams demolished. He doesn't fish, nor does he spend much time recreating on Idaho's many famous whitewater rivers. The only thing out of the ordinary in his split-level home in a quiet suburban neighborhood are two grand pianos in the front room and, in the kitchen, a six-cabinet file drawer crammed full of information on dams in the Columbia Basin.

Now in his early sixties, Burkholder grew up in Idaho in the days when salmon returned en masse to the rivers in Western Idaho. There were still quite a few around when he left Boise for college back East in the mid-1960s. When he returned in the mid-1980s to raise a family, the salmon were mostly gone. "I took my kids up to the South Fork of the Clearwater River, where I went as a kid, to show them the salmon that I remembered," Burkholder recalled, "and I took them to a spot where they always were, and there was nothing. So we went to another place. And another. Finally, I saw an Idaho Fish and Game truck. I stopped and introduced myself, explained who I was and what I was looking for. I asked this fellow: 'What happened to all the salmon? Where did they go? Was it a disease? Did we catch too many of them?' He told me, 'The problem is downstream.' I went home to my father's place here in Boise, not far from my place. He's ninety and in his old age has become something of a man of few words. I asked him what happened to all the fish. He said, 'You can have dams, or you can have salmon.'"

Based on that simple assessment, Burkholder the Younger launched his own citizen's investigation into all matters hydroelectric. He began by framing a research query for himself: what do the dams give us that makes them so important we can't have wild salmon in Idaho? He read all the relevant books he could find. He pored over government reports. He crunched the Corps' numbers. He studied the dams through the lenses of law, biology, economics. When he emerged from this undertaking in the early 1990s, the message he carried, for a short time, left him a lone voice in the wilderness: the most effective, economical way to solve fish passage problems through the gauntlet of eight dams between Idaho and the ocean was to take out four of them on the lower Snake. "My first public presentation was to the Northwest Power and Conservation Council in 1992. After that, some conservation groups, Idaho Rivers United and some others, caught wind of the numbers I'd presented," Burkholder smiled. "Nobody paid them much mind at first. But it didn't take long."

Economists, environmentalists, skeptics, and the Corps set out to vet Burkholder's numbers. Much to the Corps' chagrin, it was affirmed that his math was good. Unwittingly, Burkholder ignited a debate that's been taken up at one time or another by many of the nation's major daily newspapers, prestigious think tanks like the RAND Corporation, national conservation groups, and a herd of concerned scientists. All set about expanding on what Burkholder found. Policy experts began to amass a convincing body of economic and scientific evidence to support Snake River dam removal. This momentum reached a critical mass in 2000, when federal officials had dam removal on the short list of rational salmon management choices for the Columbia Basin. But then, as Burkholder and many others have found, rational evidence often doesn't count for much in the world of Western water politics.

The argument among the most powerful stakeholders, Burkholder found, has always been less about water or salmon and more about maintaining easy access to the cheapest electricity in the nation. Burkholder has made a second big study out of vetting the claims of these interests.

Much of the resistance to change with the Snake River dams, he found, comes from industry lobbying groups like the Pacific Northwest Waterways Association and the Northwest River Partners, which represents corporations like Weyerhaeuser, Alcoa, and Avista, as well as public utilities, barge lines, and small municipal ports along the river. A campaign called "Green Dams, Blue Skies" was jump-started by Northwest River Partners, touting the dams as a carbon-free, therefore ecologically principled source of electricity, effectively pitting one ecological virtue against the other—salmon restoration against carbon output. In late 2009, the Obama administration came to the official defense of the "Green Dams" initiative in a court of law.

When a judge inquired why the federal government proposed to end a court-ordered increase in flows on the Columbia for the benefit of salmon, their lead attorney professed his abiding concern for the health of the earth's atmosphere, one that apparently can be assuaged at the expense of the planet's water and fish. "Your honor, that [extra water] comes with a cost. And I'm not talking about financial cost. I'm talking about carbon. The more we spill," the Department of Justice's Coby Howell informed the court, "the more we are going to have to offset that with natural gas and coal."

What this budding environmental lawyer did not mention was the hydrosystem's long-standing record of service to the fossil fuel industry. Millions of gallons of gasoline and diesel are barged up and down the river every year. In the sixteen months prior to the federal defendants' stated commitment to reducing carbon emissions, five barges carrying a total of six million gallons of gas and diesel were involved in accidents on the water. One barge got stuck off the mouth of the Hood River at the height of summer steelhead migration. Another punctured a four-foot hole in its outer hull negotiating the locks at The Dalles Dam. Like much of the more controversial business conducted on the river, what really happened with these accidents is kept secret. Federal agencies, including the Coast Guard, citing confidentiality policies, aren't talking openly about what they found.

The business infrastructure of fossil fuel extraction also relies on the Corps' control of the river. Imperial Oil/Exxon Mobil and Conoco are both ferrying massive loads of mining and refining equipment up the Columbia and Snake rivers. The reason? The federal family feels obliged to help lay the groundwork for fifty years of tar sand extraction and refinement in the Kearl Oil Sands Project in Northern Alberta. From the port's terminus in Lewiston, Idaho, enormous trucks, traveling at night on desolate two-lane roads that follow several of the nation's more spectacular rivers, will complete the journey to the northern tundra, where 4.6 billion barrels of the most carbon-costly crude on earth awaits extraction.

For these and other reasons, to Burkholder, the federally adapted green dam argument is little more than spin. "These people are just not *serious* about climate change. They're just conniving to maintain their access to cheap power. Look at what's happening today with the Snake River dams," Burkholder began, consulting a Corps graph. "It's August. Supposed to be ninety-eight degrees this afternoon. What power managers would agree is a day of peak demand, without a doubt. Now the number Northwest River Partners uses to sell these four dams to the public is 1,100 [megawatts]. They'll tell you anytime they need it, they can turn on the river and get that amount of power—enough, they say, to light up three or four Seattles.

"But look at this," Burkholder said, pushing the graph in front of me. "Based on the water they have today—you can check the flows online—this is what they're getting right now: those four dams all *together* are producing only 364 megawatts of power. On a peak-load day, these power plants are running at *10 percent* of their capacity. And the power output will be minimal until next spring, when runoff starts. Now the buzzword these days is *carbon footprint*, and people will say, 'But fossil fuels produce greenhouse gas!' And they'd be right. But look at the *ecological* footprint. Does it make sense to back up 140 miles of prime river habitat, habitat for an endangered species, to produce less than 500 megawatts of electricity nine months out of the year? What kind of green energy is that?" The piano teacher paused for a moment, deftly flipping pages as if a prize student were ripping through a concerto. "Right now, this year, worldwide," he started again, "we are using 1,191,655 average megawatts of conventional thermally generated electricity. That means you're burning something, most commonly coal, to make the power. I'm betting we can replace those 'green' 364 megawatts of power without changing that fact very much. And even if we replaced all that power with coal or gas, it would be less than one one-thousandth of the output of fossil-fuel-powered electricity.

"Number one: we don't have to do it that way. Number two: if you want to do something about global warming," Burkholder summed up, tapping the 1,195,655 megawatt figure with his pencil, "I would focus on this number. I would ask Northwest River Partners what their members are doing to fight *that* number in addition to trying to save the dams. I don't see them, the Bonneville Power Administration (BPA) or anyone else calling for closing coal plants in Wyoming or the one in Centralia, south of Seattle." The reason they're not doing so, Burkholder pointed out, is that hydropower alone can't supply the power needs of the Pacific Northwest. On a hot summer day, it is coal that keeps the air-conditioning cranking in Boise and throughout much of the country. But as a half dozen corporations

that have made a killing on energy-intensive Internet commerce have discovered, hydropower is still considered green in
any renewable-energy portfolio. As long as no one like Burkholder is doing the cost accounting, signing on for long-term
megawatts of hydropower assuages the public cry to free ourselves from the dominion of King Coal. It can't hurt, either, that
this comes with one of the most generous subsidies the federal
government has to offer.

The Google corporation, whose slogan "Don't be evil" operates as a coy allusion to their socially and environmentally
conscious aspirations, has nonetheless decided that waterfront
property at the site of a former aluminum smelter on the Columbia River in The Dalles is worth its price in potential dam-
nation. Cheap electricity from the dam just upstream enticed
them. Google has spent $600 million to construct two seventy-
thousand-square-foot warehouses, each the size of a football
field, with eventual plans for a third.

In each building, row upon row of computers stacked high in
back-to-back rows and Velcroed together facilitate the instantly
gratified queries of Google searches and YouTube junkies the
world over. It's not known how many such "farms" Google operates. But some fraction of the current of the Columbia River
is now powering the stream of electronic particles that translate
into the pics of your nephew's birthday party, your brother's
favorite porn site, or your mother's online bank statement. You
can even Google "extinct Columbia salmon runs," which will
deliver, without a hint of irony, 13,100 results on the subject.
While the expediency of such an enterprise is unquestionable,
its perceived benefit to the social fabric of the Columbia Basin
is quite another matter.

Google's proposed bargain in The Dalles was delayed for six
months in 2005, while a team of do-gooders from the company
held out for an unprecedented fifteen-year tax exemption from
Oregon, sole proprietorship of a fiber-optic network built with
government funds, and, most notably, exceptional guarantees
that the cheapest price for power would be honored come hell

or high water. State and federal officials catered to Google's demands. How much power Google uses from the Columbia is now officially a trade and state secret. Lawyers and officials privy to the deal were required by Google to sign confidentiality agreements.

According to the best-educated guesses, Google is paying less than half the market rate for their juice in The Dalles. Steve Weiss, an energy analyst for the nonprofit Northwest Energy Coalition, figures Google is paying $219,000 per average annual megawatt. Thus a company that took in $16 billion in 2008 is being underwritten for $8.5 million a year in electricity costs alone, welfare furnished by other ratepayers in the region who don't get the preferential rate. This adds up to a $42,000-a-year annual subsidy for each of the two hundred local jobs that Google has created.

Devotees of expanding Web technologies speak in rapturous tones of "the cloud," the sublime term for the ever-more powerful network of computers linked together around the globe. Enlightenment derived from microchips and phone lines may one day be as voluminous and omnipresent, yet as gentle in its angelic descent to earth as condensed vapor from the oceans, falling as a gentle spring mist. But in the meantime, emissions from the cloud are far from being right as rain. Sprawling server farms like the one in The Dalles require a half watt in cooling for every watt of electricity used for processing data requests. The resulting power demand is staggering. The software security firm McAfee has figured that the electricity needed annually to transmit trillions of spam e-mails alone equals the amount required to power over two million homes, producing the same level of greenhouse gas emissions as more than three million cars. Some studies estimate the Internet will be producing 20 percent of the world's greenhouse gases by 2020.

It's been estimated Google's operation along the Columbia will use at capacity about the same amount of power as the 82,000 homes of Tacoma, Washington. That's just for starters. Amazon, Microsoft, Yahoo, and Ask.com are all making plans to build similarly brave new farms on the banks of the Columbia. In negotiating cut-rate power deals, all these companies will have whatever information they can glean from Google's haggling in their legal files and will doubtless be bargaining for as good or better terms. Google's pressing need to reduce its carbon footprint allies the company with dam proponents. According to this line of reasoning, dams are good for the environment because they don't belch carbon dioxide into the atmosphere as fast as fossil fuels do.[1] The reality, however, is not that simple. What's been selling as environmental virtue—carbon-free electricity generation—would be more accurately considered a trade of one form of degradation for another. Burkholder keeps a running list on this package of damages done by dams. Just the highlights of the list are damning enough:

- Species threatened with extinction: Snake River sockeye, spring/summer chinook, fall chinook, and steelhead. Wild coho went extinct in the Snake in the mid-1980s.
- Land lost to new reservoirs: 34,715 acres, much of it productive farm land or habitat for birds and other riparian wildlife.
- Jobs lost: thirty-nine fish canneries closed, sport fishing severely curtailed.
- Violation of Indian treaties: the Corps not only destroyed salmon runs but drowned customary hunting and fishing grounds promised to regional tribes in a treaty signed in 1855.
- Against intent of Congress: the 1968 Wild and Scenic Rivers Act, which laid the cornerstone for what would become the largest swath of reserved wildlands in the Lower 48, the Frank Church-River of No Return Wilderness, declares one of its primary purposes as the protection of salmon habitat. Five adjacent federal wilderness areas and two national recreation areas added since have created 3.3 million acres of what's supposed to function as a salmon sanctuary. But the four lower Snake dams prevent salmon from getting to the Church, or anyplace else in this vast preserve, on time.

• Degraded salmon habitat: 140 miles in the Snake River itself. But the dams also cut off access to 5,500 miles of tributary streams: the Clearwater, the middle, main, and south
forks of the Salmon, the Wenaha, the Imnaha, and the Snake
River through Hell's Canyon, to name just a few.

Nonetheless, the simple version of this equation—less carbon dioxide equals more environmental virtue—has become
the default argument for justifying the existence of status-quo
dam operations on the Columbia. It's part of a long-standing
tradition in the West of allowing moneyed interests to take for
themselves subsidies meant for a much more deserving segment
of the population.

Like the Bureau of Rec, the BPA was created to enhance
prospects for a struggling middle class. And as with its sister
agency, this mission was bent to satisfy the old Western adage
that water can always be made to flow uphill toward money.
The land and water use agencies of the nation were created to
serve the interests of the people, not corporations. Salmon are
disappearing as much because of an obstruction of justice as
obstruction of water.

The mission of the BPA is crystal clear, spelled out in the
New Deal–era law that created it. Federal power generated at
Columbia Basin dams would be sold at cost to public utilities as
they popped up around the Northwest. If ratepayers felt their
public utility was charging too much money, they reserved the
right to form a new one. The BPA would be a self-financing
agency, with borrowing power limited to the planning and construction of new projects.

Getting the electricity to market for the first few years of
the BPA's existence was no mean feat. There was simply no end
user for all the juice being produced at just two dams, Grand
Coulee in Eastern Washington and Bonneville east of Portland.
The BPA had provided for this "if you build it, they will come"
interim. In order to lure industry to the region, power could be
furnished to prospective manufacturers at below-market costs.

But here again, the law is clear: only if and when the needs of residential ratepayers were met could an industrial enterprise receive the privilege of what's known as a "Direct Service Industry" contract. Even with these DSIs, the glut of power in the sparsely populated Northwest of the 1930s gave privateers fodder for their war against government electricity. Surplus power was still a problem. Federal hydropower was in danger of becoming a monument to government overkill, an early father of the more recent Alaskan "Bridge to Nowhere." Then Japan bombed Pearl Harbor in December 1941. Suddenly, electricity and aluminum were about to change the world, though the transformation was going to be radically different from what electricity's early promoters had envisioned.

Bonneville and Grand Coulee dams became the dynamos that powered the American war machine. Aluminum was needed to make airplanes and a lot of electricity was needed to make aluminum. With Grand Coulee coming online just before America entered World War II, the Japanese had inadvertently made President Roosevelt and his dam boosters look like geniuses. Throughout the 1940s, aluminum contractors along the Columbia were using more than half of all the power offered by the BPA. The BPA, in turn, came to depend on the aluminum industry for financial justification of its existence. The effect of this grip on the region's power supply, as historian Richard White put it, "made a mockery of the project's mandate to provide the widest possible use and avoid the monopolization of electricity. . . . Aluminum had . . . hijacked the river."

Having hijacked the Columbia, the DSIs proceeded to take the river for a long joy ride that has never quite ended. As World War II drew to a close and defense demands for aluminum declined, the BPA's industrial sales team didn't change the scope or purpose of its mission. A long succession of BPA executives exited that agency and entered into cushy positions on the corporate boards of aluminum companies. In 1970, the aluminum industry gobbled up 22 billion kilowatt hours, 40 percent of the electricity sold by the BPA, consuming an energy output equivalent to five Bonneville Dams. Corporate utilities quickly secured the right to buy power at the same low rate as the public.[2] In its zeal to expand generating and transmission capacity, the agency bargained away the mandate to provide low-cost public power first and everyone else second. The BPA simply continued with its prewar mentality of persistently seeking to expand sales by creating or locating new markets, which in turn justified its relentless call for building more dams, a lobbying effort buttressed by hydropower-friendly industries, as well as by the Corps and the Bureau of Rec.

The upshot of all this federal generosity was a region of the United States that consumed electricity at a rate that would bankrupt the average Con Edison subscriber. Until some of the first conservation incentives came along in the 1970s, per capita consumption in the Pacific Northwest was three times the national average. Profligate consumption was coaxed to match prodigious production. One ad from a Seattle-area public utility featured the ideal new home, complete with an electrically heated dog house. The Western Intertie, a massive power line expansion that allowed generating stations from the Southwest to trade power with those in the Northwest, prompted the negotiation of an international treaty with Canada to build more dams there for water storage to keep the turbines spinning at full capacity the length of the Columbia. All that wasn't enough to satisfy the mindless imperative for growth.

By 1975, having dammed all the best sites on the Columbia and Snake, the BPA partnered up with a group of public and corporate utility companies to create the Washington Public Power Supply System (WPPSS). The endeavor would build three nuclear plants near the Department of Defense's Hanford facility, expanding the federal "nuclear reservation." BPA administrator Don Hodel stumped vociferously for the project. He warned of crippling power outages in the very near future and castigated those who dared voice environmental concerns: "This new environmental movement is on a collision course with the growing demand for energy. It has fallen into the hands of a small, arrogant faction which has dedicated itself to bringing our society to a halt. They are the anti-producers and anti-achievers. The doctrine they preach is that of scarcity and self-denial. I call this faction the Prophets of Shortage." Hodel himself proved to be a prophet of incompetence.[3] In 1976, using trumped-up projections for future power needs, he was such a convincing huckster that the WPPSS partners all agreed to increase the number of planned nuclear plants from three to five. He also approved an artful accounting dodge that helped the BPA around the federal prohibition against the agency buying any non-federal power. Clearing this obstacle allowed the BPA to broker the deal for the nuclear plants between itself and corporate utilities.

WPPSS would eventually become the largest case of bond default in U.S. history. Originally estimated to cost $5 billion, the final tab for the five plants, only one of which ever produced any power, was $24 billion, more than the entire worth of the Columbia River hydroelectric system. The whole fiasco left the BPA, for whom Hodel volunteered to guarantee the construction bonds, with a paralyzing, multibillion-dollar debt load, which it leaned on its regular ratepaying customers to service. Between 1979 and 1983, coinciding with a deep national recession, citizens of the Pacific Northwest opened their monthly power bills with increasing trepidation, absorbing a sixfold increase in power rates.

To head off widespread revolt, and to quell once and for all the debate between public and private ownership of utilities and power-generation facilities, Congress passed the Northwest Power Act (NPA) in 1980. It proclaimed to no small amount of fanfare that the hydrosystem forthwith would be managed with salmon and electricity production as coequals. It established the Northwest Power and Conservation Council (the NPCC, the body to which Burkholder first reported his work on the Snake dams) to pursue the goal of a steady supply of power, as well as a miraculous rebound for salmon. But private power interests made sure the new law contained perks for them, too.

To appease the privateers from corporate utilities, the NPA provided for something called a residential exchange program. Through a complicated formula, the BPA would make cash payments to these for-profit companies on behalf of their residential customers. These customers would then receive credits on their monthly bills that aggregately would equal the amount of cash doled out to private power companies. The premise was to keep public and private power concerns from beating each other senseless over who got in line first for BPA power, but the reality was that Congress had created a powerful disincentive for the creation of public utilities, a key provision in the creation of the BPA, by putting corporate and public utilities on equal footing. Though technically not a subsidy, the residential exchange program would eventually become corrupted by corporations as if it were one.[4]

Subsidies, of course, cost the BPA revenue, a fact the agency has attempted to make up for by engaging in an endless search for new markets. One boon has been the exponential growth of the metropolises of the American Southwest, where air-conditioning gets cranked just about the time northwesterners are turning down their furnaces. The BPA now sells more power to California in the spring—virtually all the power the Snake River dams produce during spring runoff—than it does to the DSIs, some $400 million worth a year. The practice of selling a block of future hydropower to any willing customer comes with the same risks as promising a fixed amount of future water to agricultural users. The exercise is speculative, based on some rather unpredictable variables like power demand on a given day or abundance of snow and rain in a given year. Critics of the BPA have long asserted that its ceaseless marketing effort to find more customers is nothing more than a nasty habit of gambling on energy futures with ratepayers' money and salmon's extinction. Heedless of such warnings, how dire a scenario the BPA could be plunged into became clear in the Enron-manufactured rolling blackouts of 2000.

As Enron was gaming the deregulated electricity market

that summer, the BPA was hemorrhaging cash in order to shore up power supplies for the constituency it had been created to serve. In a drought year, they had contracts for 11,000 megawatts of electricity service. But the river had only about 8,000 to give. After giving away its own electricity at rock-bottom prices to the DSIs, the BPA had been forced to spend $50 million on the hyperinflated electricity market to keep the lights on for its regular residential ratepaying customers. This would jeopardize the $730 million WPPSS-driven debt payment due the Treasury by the end of 2000. The solution became clear: the agency would quickly offer to buy out its subsidized DSI customers, a painfully expensive proposition that was nonetheless far cheaper than relying on the invisible hand of the marketplace, where, contrary to what the prophets of energy deregulation had promised, prices were headed for the stratosphere. To further the cause of making ends meet, they would also dump water reserved for salmon in a drought year to generate electricity, a legally dubious plan that federal regulatory agencies apparently felt was outside their jurisdiction to prevent.[5]

Meanwhile, the beneficiaries of heavily subsidized BPA power contracts were making out like bandits. The Kaiser Aluminum smelter near Spokane, for example, made $47 million in one month (December 2000), shuttering the plant, laying off workers, and reselling its BPA electricity on the open market—for $555 a megawatt hour, a federally protected 2,400 percent profit. All told, the BPA's preferred customers made a cool $1.2 billion in 2001 by reselling BPA electricity on the spot power market.

In spite of history, good reasons for hope have emerged in the form of a widely published fifty-year regional power plan, adapted by the NPCC. Based on some cautious forecasts, the plan recognizes conservation as the biggest energy asset the region can deploy. All the future energy requirements of the Pacific Northwest over the next half century—some 6,000 megawatts—can be met though conserving energy, which can be accomplished through technology instead of sacrifice. But implementing the plan will require some political and social change to the status quo. The original NPA recognized conservation as a legitimate source of energy supply. Since then, according to the NPCC, 3,600 megawatts of power already have been saved through conservation measures. According to the law, these savings were supposed to be credited to the NPCC's Columbia River Basin Fish and Wildlife Program, in anticipation of a more equitable share of the river's flow being dedicated to salmon restoration. The water it took to generate those megawatts, in other words, could be dedicated to letting the river run more like a river. Instead, the BPA has absorbed these savings into keeping its wholesale rates at half the market price. If the BPA gets its way, ratepayers and taxpayers will end up paying for this theft twice. In crunching its numbers for the cost of replacing energy lost to the removal of the Snake River dams, the BPA calculates these costs at current market rates.

Burkholder, of course, knows this history, and the resulting predicament, as well as anyone. Overcoming entrenched corporate and political interests does not strike him as an insurmountable task. Prevailing on such matters, Burkholder believes, can begin with the smallest gesture. "We are addicted to fossil fuel," he says. "We burn coal in Wyoming to keep our houses cool in Boise. That's amoral. And since I know that, I don't run an air conditioner in my house. It's not much, but that is directly addressing global warming. Encouraging companies like Google to switch to hydropower is not solving the problem. These dams are not providing clean energy because they directly cause species extinction. Even coal-fired energy plants don't do that. The dams are basically a 140-mile-long strip mine, except there are no reclamation laws. The package of negative impacts is much more direct and potentially bigger than dirty coal."

It's quite a distance, Burkholder acknowledged, from turning off the air-conditioning to changing the way the world sees energy: less as a commodity than as an input from which many other good things might grow. Or as simply a basic requirement of living, worthy therefore of our most precautionary, respectful, and wise decisions. But Burkholder is undeterred. "I look at it this way," he reflected. "I've been crunching these numbers for fifteen years now. Each time I do it, it seems to make my original points stronger. As far as I'm concerned, I won this argument fifteen years ago. The rest of the fight is simply getting the information out there. I look at things a different way. I teach kids to play the piano. So, I'm accustomed to taking a kid who knows nothing and turning him into a piano player. I'm familiar with what it takes to go from nothing to something. It can be a long process. But that is why it is cosmically important that we talk to each other. We need to keep communicating. The next convert is just a conversation away. Information trickles up to my congressman, and he can't ignore it.

"From my point of view, as a person who teaches, I still see terrific hope. As long as you can talk to someone calmly, show them the facts, you can bring people along," Burkholder serenely opined. "But it is time to bring ethical and even aesthetic arguments back into the public arena. I've kind of retired from this project. At least, I don't work on it six hours a day like I did up until a few years back. I still talk to school groups and church groups, business groups, whoever asks. The numbers are the easy part to talk about in public. But salmon have a value beyond the dollar. There's an argument to be made for beauty. It drives a sense of the communal, what we all share. And that makes it a moral argument as well."

I bid Burkholder farewell and drove north up the Payette River, on over into the mile-high Stanley Basin, the alpenglow on the Sawtooths making them appear like the canines of some gargantuan monster blowing fire out of its subterranean belly. Over the divide between the Payette and Salmon rivers, I stopped in the dark and rolled out a sleeping bag on the banks of Marsh Creek. I knew there would be a few salmon spawning in the fall, strong wild fish that would have leaped the equivalent of a three-story building at Dagger Falls after having swum a distance equivalent to driving from Portland to San Francisco.

I also knew that all the arsenal a person would ever need to make the kind of arguments in defense of beauty that Burkholder advocated could be found by launching on Marsh Creek and touring through three hundred miles of one of the most spectacular wilderness river systems on the planet.

That a piano teacher could wind up as the unlikely pioneer of lower Snake dam removal is part of a lineage that started with keeping the last hundred of those three hundred miles free-flowing. It was a victory even less likely than Reed's.

6

HOW THE MIGHTY WERE FELLED

*Environmentalists may be hell to live with,
but they make great ancestors.*

—ANDY KERR, environmental consultant

Landscapes harbor secrets. Cartography can help reveal these terrestrial codes. A map of the two main rivers of the Columbia Basin looks roughly like a hastily squiggled wishbone, the main Columbia forming the northern half of a prostrated Y, the Snake the southern half. For geographical reasons, the wishbone is not symmetrical. For political reasons, the southern half of the wishbone doesn't mirror the northern half in the radical transformation of its hydrology. The Snake is missing a key feature for making a river basin work as an efficient machine. The maps show it's short one big lake.

The way to make a river system pay the highest possible dividend under the management philosophy of "Total Use" hinges on the presence of a large storage reservoir in the upper basin that can be used to regulate the flow of the entire system below it. In addition to being the largest hydroelectric plant in the country, this is the service that Grand Coulee performs. All the dams downstream—Chief Joseph, Wells, Rocky Reach, Rock Island, Wanapum, Priest Rapids, McNary, John Day, The Dalles, and Bonneville—have water to spin turbines year-round in large measure thanks to Grand Coulee. But the logical location for a similar facility on the Snake is off-limits, protected by a seven-hundred-thousand-acre refuge that forever excludes dam-building from its list of officially sanctioned activities.

In the mid-1960s, the archdruid himself, David Brower, hired a bright-eyed young lawyer from Ohio to represent the Sierra Club in the Pacific Northwest. Brower informed his new hire that his territory was the Pacific Northwestern region of the *continent*, not the country; as such, he'd be responsible for work from the Yukon to Yellowstone. Moreover, there wasn't a lot of reserve in the travel budget. The aspiring attorney took the job anyway, moved to Seattle, promptly took a long weekend off, drove with his wife to Hell's Canyon, and fell in love with the place. He didn't know it at the time, but this was the first step in a fight that would bring the era of big dam building in the United States to an abrupt end.

This was accomplished by successfully contesting the fifth dam that was supposed to go in on the lower Snake. It would have drowned the last remaining wild portion of the river as it flows through Hell's Canyon. It would have been the Snake River sister of Grand Coulee Dam. It would have ruined a lot more salmon habitat, further muting the argument for the Snake as the last best hope for salmon recovery. It would have meant another river basin wholly given over to the machinery of "Total Use." And it would have had a funny name: the High Mountain Sheep Dam.

That attorney Brower hired was, and still is, named Brock Evans. Now semiretired and living in Washington, D.C., Evans remembers the three basic elements of the epic battle against the fifth dam. The first was how powerful and well-connected the opposition was. The second was how futile the fight looked. The third was how simultaneously giddy and anxiety-ridden he was, a thirty-year-old rookie attorney operating on a shoestring budget, going up against the self-assured wealth and experience of government and corporate lawyers. "It's hard for me these days even to relate how hopeless our case seemed," Evans told me by phone from his home in D.C. "Dams were the accepted wisdom. Any notion of a 'wild river' was just heresy. There were no environmental laws. Not a single one. No guidance or any precedent. Not only that, there was no constituency. Not a single senator or congressman backing us. The only thing we had were a few local groups, who were very courageous given the open hostility toward us. We knew we couldn't stop the dam. The only hope was to slow it down and hope for a law or some other kind of legislation that would have the power to kill it."

Powerful as they were, the chink in the armor of "Total Use" water fanatics was that they never really stopped competing against one another for the upper hand until it was too late. The post–World War II contest between New Deal Democrats and Square Deal Republicans has faded from public memory, but in the days of Presidents Truman and Eisenhower, Hell's Canyon became the main battlefront in a war between the federalizers and privateers. For many years, the largest crowd ever assembled in Idaho was when Eisenhower came to Boise, in 1952, to announce before twenty thousand people that the government was getting out of the dam-building business. To Total Use advocates, those days now look like a missed opportunity, something akin to the halcyon days of the American auto industry, when General Motors was so big it could have simply bought up all its competitors. Combining forces when it mattered most, the privateers and federalizers might have succeeded in completely locking up the entire Columbia Basin. Instead, on the Snake River, they locked horns so intensely they didn't notice the emergence of Evans' ragtag group of underdog environmentalists tripping them up.

Four private utility companies combined forces and applied jointly to the Federal Power Commission for a license to build the 670-foot-tall High Mountain Sheep Dam, a mile upstream of the mouth of the Snake's biggest, wildest tributary, the Salmon River. In issuing the license, the Power Commission sided with the privateers in a bitter, divisive battle with the Corps and the Bureau of Rec. The feds had pitched their own project for Hell's Canyon. Dubbed the Hell's Canyon High Dam, it would have been built downstream of the mouth of the Salmon, cutting off access to almost all of the Idaho wilderness for chinook and steelhead. Three times the bulk of High Mountain Sheep, it was to be the Bureau of Rec's crowning twentieth-century achievement. A reservoir ninety-three miles long would have stored four million acre-feet of water. Some of the electricity generated at the dam would have gone to power massive pumps that would have sent water east through tunnels bored into the mountains separating Southern and Central Idaho. Combining forces with existing Bureau of Rec projects on the Boise and Payette rivers, the canal would have transformed four hundred thousand acres of desert around Boise into farms, ranches, and subdivisions. The privateers' proposals eventually won the license, based partly on their well-publicized claim that only corporations, with their ultimate responsibility to shareholders, could build a dam on a scale commensurate with reasonable public need. Without a national depression or a world war to justify the massive scale of the public works projects of the New Deal, the privateers' government overkill argument won the day, with the Power Commission issuing the license for High Mountain Sheep in early 1964.

By 1967, Evans' concern for Hell's Canyon had gotten personal. "I loved the place," he reminisced in his memoirs. "From the first time I had journeyed through the Canyon, enchanted by its majesty and beauty, it was as if some old lost chord had been plucked inside. My heart sang to a new kind of music I hadn't even known was there. I loved Hell's Canyon, and vowed to give everything in my power to try to save it." To that end, in August 1967, on behalf of the Sierra Club and a fledgling group that would come to be known as the Hell's Canyon Preservation Council (HCPC), Evans filed formal protest with the Power Commission, arguing the river should be allowed to run free. For the dam builders, this was pouring salt in an open wound. Earlier that summer, the Department of the Interior won a lawsuit against the Power Commission. Based on their disastrous experience in dealing with Idaho Power's Oxbow and Brownlee Dams on the Snake just upstream, the Interior saw the new dam proposal as a potential fish-killing machine and was not anxious to be the federal agency responsible for mitigating the damage. Biologists in the Fish and Wildlife Service already knew there was nothing that could fix the Idaho Power dams. Passage for fall chinook and steelhead consisted of an inept attempt to trap adult fish in a net and haul them above the dams. Idaho Power quickly gave up on any form of fish passage, in clear violation of the license that had been granted by the Power Commission. Instead, it negotiated to contribute funds to a fish hatchery program. The biological negatives of hatcheries had not yet been widely proselytized, but the political problem with them was already obvious. Hatchery programs reliably wind up putting fish not back where they came from but far downriver in the lower Columbia, closer to urban sport and coastal commercial fisheries, becoming another means by which dams transfer wealth downstream.

The U.S. Supreme Court agreed to consider Evans' complaint. The question before the Court was whether such gestures, and more importantly, the dam itself could be seen as serving the public interest. Justice William O. Douglas wrote the majority opinion directing the Power Commission to weigh "the need to destroy the river as a waterway, the desirability of its demise" and to reconsider the license.

Suddenly, Evans' hopeless mission was crystalized into two clear goals: organize a campaign to convince the public that Hell's Canyon's demise was not desirable, and convince the Power Commission not to reissue that license. Neither task seemed likely to meet with anything resembling success. To begin with, the Power Commission began its rehearing over the High Mountain Sheep license in Washington, D.C., but neither Evans nor the Sierra Club could afford regular travel to the nation's capital. Evans petitioned for public as well as legal hearings in the Northwest and, in the process, gained what he termed "a closet ally." The professional staff at the Power Commission surreptitiously wanted to help with a nascent wild rivers campaign. Still, the deck was stacked against him. "The power companies were littering the hearings with these glossy 'Snake River: The River That Wants to Work' pamphlets," Evans remembered as we spoke. "They had ads on television. Their attorneys tried to have me banned from the legal hearings, telling a judge that based on what they'd seen from my organizing and public relations work, I was 'too political.' I explained to the judge I was the only one who could do what was needed. Then, too, we had to be credible on the power issue. Large tomes prepared by paid experts were produced by the companies to prove that if the Northwest didn't get electricity from this dam, the whole economy would collapse. There was no point in arguing for conservation back then." To appease the conventional wisdom of the day, the Sierra Club openly advocated for nuclear power in those years.

The public hearings were held in Lewiston and Portland. Evans spent most of the summer of 1968 driving around the Palouse, Eastern Oregon, and Idaho. He felt his side could hold its own in Portland but was less sure about the Lewiston hearing. However, Evans said it was at that hearing that he was relieved to sense a rising political tide against the completion of a fifth dam: "The HCPC did outstanding organizing, speaking, and publicity work. Then we went to Portland, and we just smashed the other team."

The other team, however, decided it was time to combine rosters. Secretary of the Interior Stewart Udall, head of the department that successfully sued the Power Commission just a few years earlier, announced suddenly that the feds were in favor of a dam in the Hell's Canyon, albeit a scaled-down version of the original plan at a different site. It was also announced that the thirty-years' war between public and private factions had been brought to a truce: they would pursue a jointly owned license from the Power Commission.

Then, two politicians whose careers would later be tarnished by scandal intervened on behalf of saving the wild Snake. Bob Packwood was elected from Oregon to the Senate and made it known he wanted a part in saving the canyon. Meanwhile,

Evans flew to Washington, D.C., to convince the new Nixon administration to reverse the Interior's new pro-dam stance. The visit turned out to be with Nixon aide John Ehrlichman. "I am a fanatic environmentalist, Brock," Ehrlichman confessed. He finagled a meeting with the Interior's second-in-command, who came out against the dam.

The roller-coaster ride continued. As part of the first-ever Earth Day hoopla (April 22, 1970), an NBC film crew, Packwood, and a pack of free river advocates floated through Hell's Canyon. The ensuing *Nightly News* special report earned a slot in prime time, inciting a wave of national protest against the dam. Packwood introduced a bill to create the Hell's Canyon National Recreation Area, an idea fleshed out during the float through the canyon.

In 1972, the federal court hearing the Power Commission's reapplication process reissued the license for a public-private dam in Hell's Canyon. No one was especially surprised by the decision. But the judge, cognizant of the brewing controversy, delayed the new start date until 1975. The race was on: if Packwood's bill, or some version of it, passed through Congress unscathed before the end of 1975, no new dams would be built—forever—in Hell's Canyon. On November 16, 1975, President Ford made it official. Hell's Canyon of the Snake was off-limits. Wilderness areas were designated on the crest of either side of the canyon and a final, fallback third attempt to locate a dam on the Snake was deauthorized. (This dam would have been less than twenty miles upstream of Lewiston at Asotin, Washington.)

"We did it," Evans proclaimed, the satisfaction still palpable in his voice thirty-five years after the fact. "And if I could point to any one legacy of the decision, it would be to point out that this was another key place in the beginning of the modern wilderness movement in the country. Look at all those green spots on the map, all the wilderness areas and national recreation areas we've set aside in the years since. Every one of those green spots has a story to tell, a battle that's worth remembering."

He recalled with a chuckle the misplaced disdain with which he greeted the latest "Working Snake River" campaign, thinking dam boosters might have dusted off the phrases from those pamphlets he found everywhere in the summer of 1968. To his delight, he heard instead that Save Our Wild Salmon had co-opted the phrase from his former adversaries as part of the effort to tear down the dams on the lower river. "I think this younger generation could really do this," he said. "Now they have laws and a constituency on their side. That lower canyon was just a magnificent, lovely, solitary place. I'm happy to have lived this long to at least see the possibility that other people might get to see it again."

Another legacy of the battle for Hell's Canyon, then, is a possibility as equally generous as wilderness officially inventoried and set aside: the fight to rewild a river once stilled. With its integrity intact, the land and its rivers still hold secrets. But it's the eyes and ears of people that do the translation work. By the turn of the twentieth-first century, more people were hearing what the river had to say. Without that giant Grand Coulee–scale plug at the head of a healthy hydrosystem, the rest of the Snake River dams might not be worth keeping.

7

WHEN THE LEVEE BREAKS

Mean ol' Levee, taught me to weep and moan/
Got what it takes to make a mountain man leave his home.

—KANSAS JOE MCCOY AND MEMPHIS MINNIE

Dams have displaced towns, destroyed jobs, inundated land,
and extirpated salmon from their home rivers. But their power
to destroy goes well beyond the toll taken on quaint riverside
villages and fish. On the lower Snake River, they drowned history. In the early 1960s, while the Corps was scurrying to realize
its vision of bringing the lower Snake River under total human
control, Dr. Richard Daugherty of Washington State University's Department of Archaeology, working right under the Corps'
nose, was pursuing a vision of a more complete record of human history. So fervent was Daugherty in this pursuit that, quite
ahead of his time, he saw that the most perspicacious deciphering
of archaeological clues would require a department faculty with
a variety of specialized expertise. The result was an interdisciplinary graduate program unlike any other in the world at the
time. It soon attracted some of the country's brightest graduate
students.

Meanwhile, a local rancher named John McGregor had been
bugging Daugherty about a shallow cave near the confluence of
the Palouse and Snake rivers on the property of a neighbor
of his, one Roland Marmes. Daugherty thought the site promis-
ing enough that he paid Marmes to dig a trench there. Hiring
Marmes' bulldozer turned out to be well worth the money.

The Marmes Rockshelter, near the confluence of the Palouse and Snake rivers, is one of the most important archeological sites in the Pacific Northwest. At the time of its discovery in 1965, it was the most important in the world. Archeologists discovered a place that had been used for shelter, storage, and human burial for eleven thousand years. Methodical excavation produced shell beads, bear teeth, and, in one case, the hoop from an infant's cradleboard. Marmes Man, the skeletal remains of one unearthed specimen, was found to be ten thousand years old, seven hundred years older, give or take a century, than his Johnny-come-lately neighbor discovered downriver in 1996, Kennewick Man. At the time, Marmes Man's bones were the oldest known in the Western world.

But the Corps had no intention of letting nature or science run its course. They had already condemned the Marmes' land to make room for the future. Lower Monumental Dam, the second of four dams to advance up the lower Snake River, was on the fast track to completion. Not even the oldest signs of civilization in the Western Hemisphere could keep the Corps from seeing its mission through. Marmes would have to be flooded.

Soon, Senator Warren Magnuson (Washington) intervened on the scientists' behalf, and a compromise was brokered. The Corps begrudgingly agreed to a year for further exploration. And at Magnuson's behest, the Corps would protect Marmes.[1] A cofferdam would be built around the site, protecting it from the reservoir's rising tide. The meticulous work of dusting off the secrets of the ages could continue unabated. But the Corps, so unfalteringly competent when it came to flooding the desert, proved to be inept at keeping even a small portion of it dry when the need arose. The cofferdam failed, and Marmes was inundated with forty feet of water.

All too eager to bury the past, the Corps promised a future that has never arrived. No one has gotten rich from these dams, nor did anyone foresee that one of them would instead court a brewing ecological catastrophe. If the past is a remembrance and the future a promise, the present civilization along the lower Snake sits on a timeline sandwiched between a dam-induced amnesia and a river-borne dread. Far from being resigned to whatever fate is handed them, a small group of citizens is working as diligently as those Washington State professors did forty years ago to avoid another inept rendition of the Corps' idea of flood protection.

The present-day trouble is silt. Much of the Snake River watershed drains a region of intensely irrigated agricultural lands where loss of topsoil is an ongoing, and to a grand extent unmet, challenge. A lot of this dirt is discharged into the Snake, where, thanks to the reengineered river, it settles out in a bad spot. Twenty miles above Lower Granite, the most upriver of the four dams, around Lewiston, Idaho, the water of the Snake begins to slow, signaling the end of the river's brief appearance as a free-flowing river through Hell's Canyon and the resumption of its life as a series of reservoirs. Lewiston sits on the banks of the river where it makes this transition. The Snake flanks the west side of town, while another big river, the Clearwater, flanks the north side; the two big rivers join on the west end of downtown Lewiston, narrowing the city to a prominent point overlooking their confluence. Situated in the midst of two unruly rivers, the city has always known that flooding was a risk. But for the past thirty-five years, the Lower Granite Dam has been slowly transforming this risk into an eventual certainty.

Silt in the Snake is piling up at Lewiston faster than anyone cared to consider when the dams were built. The river has been faithfully delivering more than a million cubic yards of silt annually since the Lower Granite was completed in 1975. The river through Lewiston has become shallower, on average, at the rate of more than an inch per year. Pilots maneuvering the big barges on the Snake were the first to reckon with this modern problem. At least one unlucky steelhead fisherman was a close second.

Steve Pettit, a retired Idaho Department of Fish and Game biologist, spent his career studying methods by which salmon might pass through the eight dams between the ocean and

Idaho waters. As the science behind fish passage became better informed, his was a job that found him increasingly at odds
with the Corps of Engineers. In the last decade of his career, his
opinions frequently clashed with the prevailing political views
from the governor's office as well. In the mid-1990s, Idaho
elected conservative Dirk Kempthorne as governor. The state's
Office of Species Preservation was moved from the Fish and
Game headquarters to the governor's mansion. Kempthorne,
who would later serve as President Bush's secretary of the interior, didn't care much for enlightened scientific opinion nor
democratic debate on such matters and unequivocally sided
with pro-dam forces against endangered fish. On several occasions, Kempthorne would send a staffer to sabotage high-level
salmon meetings where Pettit presented scientific findings.
"I would get done speaking," Pettit remembered, "and then
someone from the governor's office would jump in and say, 'We
would like you to disregard what Mr. Pettit is saying, he does
not represent the state's position on this issue.'" After retiring in
2005, Pettit, an accomplished fly fisherman, sought solace casting for steelhead in the Clearwater River, which runs a stone's
throw from his front door. Hip deep in a good run one December morning, he saw the winter-low water of his home river
surge suddenly. The precarious bit of wading that had gotten
him to a favored spot became the Christmas swim of his life. "I
knew what had happened right away," Pettit recalled. "Another
barge had gotten stuck down in Lewiston, and the port called
the Corps for an emergency release of water from Dworshak [a
dam on the Clearwater River upstream] to get it unstuck. But
the thought did occur to me," wisecracked Pettit, recalling his
skirmishes with dam proponents, "that maybe the other side
was getting the last laugh."

Stranded barges and body-surfing fly fishermen are bad
enough, but the worst predicament the silt has brought down
is on the levees that protect Lewiston. Once a town hemmed in
on the banks of two of the continent's more spectacular rivers,
Lewiston now lies several feet below their waters, thanks to the

reservoir behind Lower Granite Dam. Lewiston would be as lost as the civilization at Marmes if not for the Corps-built levees that have thus far kept it dry. But with the amount of sediment building annually, the Corps itself has begun to deliberate over how long the levees can be reasonably expected to hold.

Technical discussions of a levee's effectiveness borrow from a shipwright's vernacular. *Freeboard* on a boat means the height of the outside of the hull from waterline to gunwale. On a levee, *freeboard* denotes the height above what the Corps terms "standard project flood," a theoretical worst-case scenario for high water. The freeboard on the Lewiston levees was calibrated to be "standard project flood" plus five feet, a margin that engineers believed would spare the city from all but the most biblical of deluges. This margin of error has shrunk from the river bottom up every year since the dam was christened. Subtract the three-and-a-half feet the riverbed has risen from the once ample freeboard the Corps planned, and Lewiston now has only a foot and a half of levee to spare. With the levee shrunk from wall to curb, new silt will shrink the available freeboard to a speed bump with the passage of time. Every year, in other words, the odds inexorably increase that Lewiston will be inundated by the dam that was supposed to be its salvation.

The Corps now admits that even a ten-year flood, an altogether foreseeable scenario, would put Lewiston at risk. The announcement was greeted initially in the politically conservative air around Lewiston with a kind of cognitive dissonance. Conventional wisdom had it that everyone wanted their meager salmon runs, their dams, and their levees all in tact, no compromises; as usual, it was the Corps' job to figure how to make it all work. The Corps, for its part, stuck by its preferred solution: furious dredging, coupled with a feasibility study on raising the levees a few feet, would see the town through, at least for another decade or two. This fix would have its complications as well. Dredging is expensive. Minimum cost is pegged at $2.7 million a year and could run as high as $36 million annually, an expense the Corps would like taxpayers to count on for the next fifty years. But scooping and stirring the sediment compromises water quality, clouding the river and resuspending material that contains concentrated pollutants from agricultural, industrial, and mining waste.

The Corps' plan to dredge like crazy is not new. The agency had signed a procedural document called a Record of Decision in 2002 to take on the increased dredging. But a consortium of environmental groups successfully sued the Corps to stop the plan while an environmental impact study is completed. The plaintiffs won this round; the Corps was forced back to the drawing board and is now studying the situation and drafting a new plan, due in 2012, to meet the court's standards.

All this has taken place without the Corps admitting a key point: dredging even with reckless abandon will not keep Lewiston ahead of the silt. Forty to fifty million cubic yards of it has been laid down on the riverbed; merely keeping pace with the annual deposit would require about fifty thousand standard-size dump-truck loads a year, along with an invasion of heavy equipment and infrastructure that would have parts of Lewiston looking like a giant aquatic mining operation.[2]

In the meantime, the Corps is undertaking something it calls with a straight face the Programmatic Sediment Management Plan. Quite ambitiously, the Corps hopes to inventory sources of silt around the Snake River Basin—all 108,000 square miles of it—and begin the process of persuading farmers, ranchers, developers, city managers, and foresters from Jackson Hole to Asotin to implement soil-conservation practices that might alleviate the sediment troubles at Lower Granite. Conservationists point out that, while the Corps' intent is certainly praiseworthy, not only is such an undertaking unlikely to make a difference in the near future, when it will matter most to Lewiston, but that such a watershed-wide initiative is outside the Corps' jurisdiction. (The Corps has no authority outside water and land directly affected by its projects.)

The other possibility for a temporary solution is to raise the levees. But this would add to their already imposing height above the river, would be a dramatic and wildly expensive intrusion to the town, and would carry a minimum price tag of $95 million. The history of the levees' construction will make that option a tough sell.

When the plans to dam the lower Snake were sold to the region, many folks around Lewiston objected to being visually cut off from the river by the levees, a penalty the Corps mitigated (only after loud and persistent protests)[3] by liberally sprinkling them with grassy parks, picnic shelters, boat launches, and a bike path. Raising the elevation of the riverbank again would require a radical restructuring of these amenities and, likely, a long interval in which access would be severely limited. Worse, this extravagant interruption would at most buy another decade's reprieve before the river caught up to the added freeboard.

Though not yet a sea change, attitudes about the dams around Lewiston are shifting. Dams were sold in the 1950s and 1960s as a surefire means to prosperity, a cornucopia of electricity, irrigation, and navigational benefits that would lead to agricultural and fiscal bounty. A half century later, Lewiston must reckon with the possibility that one of these dams might wreck the town instead. Further, the big fix would require Lewistonites' acquiescence yet again to the Corps, allowing this engineering wing of the military to preempt the river's havoc with more man-made chaos of their own, altering dikes and levees, railroads, streets, and bridges, and destroying a greenway that many locals identify with as pridefully as some Bostonians do the Fens. Getting rid of the dams that make the levees a necessity, is looking more palatable than it did a few years ago. Whether or not that option prevails depends on a lot of people for whom river restoration, usually defined region-wide by the effort to bring back salmon, has never been a top priority.

Jim Kluss has lived in Lewiston all his life. He owns an appliance store downtown and sits on the city council. Kluss used to enjoy cruising the river through town. "We didn't even put our boat in the water last summer," he recollected to me one morning in his shop, "but the summer before that we did, and the marina had to mark the way out to the river." Without a depth finder and a deft touch at the wheel, he recalled, it was almost impossible to get out to the main channel without churning the boat's propeller in muck. Kluss, like many Lewiston residents, is not an outspoken advocate for dam removal. He's suspicious of claims from conservation groups that removing the dams would be the shot in the arm that they were promised when the dams were built. But Kluss does remember his city before the dams: "I would say, as somebody growing up here, that from when the dams were proposed, supposedly with all the economic benefits the dams were going to bring for us, that they've fallen far short of that. And I really think very few people would disagree." One of the larger hindrances in overcoming the impasse the levees have created, according to Kluss, is that the Corps locally doesn't operate as an agency that lives up to its national motto: "Relevant, Ready, Responsive, Reliable." Living in a small, relatively isolated town makes it more of a challenge to gain the ear of the Corps. "If the levees are overtopped, a lot of people are going to be watching CNN and asking 'Where's Lewiston?'" Kluss told me half-jokingly. "I told the paper yesterday, I think sometimes we are just gathering statistics so that when Lewiston does flood at least some of us in city government will be able to say 'I told you so.' But the vast majority of people here don't know about what's going on with the levees. I think that at least partly comes out of a sense that the Corps of Engineers hasn't or isn't going to respond to what people here want or need."

Beyond the collective sense that the Corps will do whatever the Corps wants, at least some of the apathy derives from the fact that the dams became such an economic dud. "This would be quite a different discussion if what the dam boosters promised had come true," Kluss said. "People would be saying: 'Tear the dams out? Are you crazy? Look what they've done for this place.' But no one got rich, and the choice we have now is dredging, raising the levees, or tearing out the dams. And I can tell you, there's little to no support around here for raising the levees. We'll see what happens with the dredging. If that doesn't come to pass, you might see a lot of folks around here start lining up behind dam removal."

The trade-off for being separated from the river was supposed to have been the Port of Lewiston, proudly advertised around town as "Idaho's Only Seaport." A modern port suggested images of burly longshoremen, stacks of pallets, lines of trucks, and a battalion of cranes and forklifts, a place bustling with the work of international trade. The current Port of Lewiston presents a slightly different scenario. It directly employs, at any given time, only twenty to twenty-five people. The steady stream of ships is conspicuously absent, as are the lines of trucks and the loaded docks, as if the longshoremen's union had commenced a strike, and the picketers had gone for a long lunch. The operation does an anemic business mostly shipping grain to the larger Columbia River ports. And what it does ship is wholly dependent on the Corps' dredging of a 465-mile-long shipping channel from Astoria, Oregon, near the Pacific Ocean, all the way to Lewiston, a subsidy worth $25 million a year. Lewiston is also dependent to an almost exclusive degree on Portland's port, since most of what leaves Lewiston by water winds up there.[4] The Port of Lewiston has never been financially solvent, let alone become a source of revenue for either Lewiston or the counties that surround it. In fact, residents of Nez Perce County, where Lewiston sits, are taxed to further subsidize the meager business the port conducts. The port authority, along with the local property tax, was created in 1958, anticipating the arrival of slackwater all the way to Lewiston. Residents were guaranteed the tax was a temporary measure and that by 1968 the port would be self-supporting. Fifty years later, the tax is still being levied. But even with federal and local subsidies, nobody at the Port of Lewiston is getting too fat: business at the port was off by two-thirds in mid-2008, due to a somewhat disappointing wheat crop. In July and August of that year, the port took in a paltry $46,000.

This dismal news comes atop massive bills coming due for

upkeep on the dams, locks, and shipping channel, which comprises the generous federal outlay the barge business relies upon to float up and down the river. According to the Corps' own numbers, simple maintenance and repairs for the Columbia and Snake rivers navigation system through 2015 could easily exceed $500 million. This excludes the bills for major repairs like replacing a damaged lock downstream at John Day Dam for a cool billion.

The barges that run up and down the lower Snake exist alongside rail lines and highways that served the region's needs quite well until 1975, when it became possible to float the year's wheat crop to Portland rather than sending it by truck or train. I recollect something Kluss told me, that the area around Lewiston has not grown substantially in forty years. I wonder if roads and rails would still suffice if the dams were gone and the salmon given a chance to recolonize a free-flowing river.

This is not a line of thinking that's especially welcome to David Doeringsfeld, the Port of Lewiston manager, who sees the river much the way that dam boosters envisioned it fifty years ago. "We've done everything we can for these fish," Doeringsfeld told me in his office one afternoon. "And we've spent billions of dollars, and they're still disappearing. The question to me, then, is how much more do we want to spend? These dams were built for the public good. Do you really want to take away that good to gamble on improving salmon populations by just a few percent?" When I uttered the phrase "endangered species" in the context of a question about the historic low numbers of salmon, Doeringsfeld's hackles practically bristled. "Who says there are that many fewer fish than there used to be?" he asked. "The fact is that no one knows how many were here before the dams. Do you have proof there were more fish?" I shrugged my shoulders. "I'm waiting," he smiled, fingers tapping smugly on his desk.

I hadn't encountered this kind of fatuousness yet in the vitriol of the salmon wars. But I hadn't come to the Lewiston Port Office to debate. What I'd really come to ask about was the silt problem. Doeringsfeld maintained that silt is merely a bureaucratic issue. "The Corps of Engineers gets their money in different categories," he told me. "They've got red money for emergencies, blue money for maintenance, and what I've heard is that it's just a matter of getting the money into the right column."

But even if the money is in the right column, I pointed out, the Corps has thus far been granted legal permission to dredge only to maintain the shipping channel. The silt settling out at the confluence of the Clearwater and Snake is not covered by this permit, a point that environmental groups gained in winning a legal battle against the Corps to stop its dredging. Doeringsfeld leaned into his desk as if letting me in on a seldom-whispered secret. "What they're doing," he started, speaking of the plaintiffs in that case, "is obstruction, pure and simple. They make it so that the Corps can't dredge, and suddenly they think they have another point in their favor for what I think—and what most people still think around here—is a totally unreasonable position: taking out the dams. This is where the Endangered Species Act (ESA) and whatever other kinds of laws govern these cases should be changed. Make the plaintiffs liable for court costs when they lose, make 'em liable for economic losses as well." The port manager paused pensively for a moment. "Here's a story that will never get written," he said. "When and if the flood hits Lewiston, it'll be the fault of these environmental groups as much as anyone else. And I bet they won't be here filling sandbags or cleaning up afterward either."

Throughout the course of our interview, I couldn't help but feel that Doeringsfeld seemed to regard me with an initial skepticism that blossomed quickly into a full-blown suspicion, a thinly veiled disdain derived from deep mistrust. His surly demeanor, his open scorn for environmentalists, made me wonder if he was interested at all in any kind of collaborative solution. So I made an appeal that had saved me before in my inquiries among these conservative heartland types, the case for local control. Self-determination. Independence. Forty acres. A mule. Lewiston for Lewistonites. That sort of thing.

"Don't you think," I queried Doeringsfeld earnestly if not somewhat tentatively, "that maybe the best solution here is for the locals to do their best to put their differences aside and maybe try to come up with some sort of united front on the dam-and-salmon issue?" I mentioned coalitions of tree-hugging conservationists and gas-huffing snowmobilers in Western Montana that were putting their best foot forward to the Forest Service and Congress, one party acquiescing to more wilderness, the other to a secure timber supply and designated roads for their arctic crotch rockets. "It just seems odd to me," I said, self-consciously rubbing the scruff on my chin, "that people in a small town, where part of the appeal is knowing your neighbors, wouldn't at least try to protect whatever interests they found they had in common."

And it was here that Doeringsfeld revealed to me the lasting and most damaging curse of the dams that make his job possible. "That idea around here," he began, smirking at me for the first time with the incredulity a hardworking, patriotic heartlander will sometimes regard a slicker from the city or a slacker from the coast, "wouldn't do us any good even if we could all get to the same page. All of these decisions are made in courtrooms and law offices and judges' chambers in Seattle and San Francisco.

"There's no control around here over *any* of it," he declared, emphasizing his point with the same gesture an umpire uses to call a base runner safe. "I used to spend about a third of my time on the salmon issue, meeting with the tribes, meeting with guides, but not anymore. Ultimately, what anyone thinks around here is not going to make a bit of difference. There's too much money and too much power at stake."

We made a little more small talk—Doeringsfeld is hopeful about a projected rise in shipping containers coming by truck and leaving by boat at his port—and then I left his office for the noonday sun, a stroll along the river, and later, an apparently much needed shave.

As I walked the bike paths atop the Lewiston levees, watching the rhythmically twitching rods of hopeful fishermen plying the river, a screeching gull bombed the ribbon of asphalt before me with an audible splat. Doeringsfeld deemed the people of the Palouse to have all the political pluck of the citizens of some thoroughly trampled banana republic. What bothered me as much as dodging the seagull sorties that afternoon was that he was in some ways right on the money in making this judgment.

Lewiston is up against the Corps, which employs a few thousand more people than Lewiston has residents (about thirty-five thousand), a budget of around $12 billion a year, and ongoing projects all over the globe, including the rebuilding of the oilfields in Iraq. As the last job helps to illustrate, the Corps gets its projects budgeted by aligning itself with powerful congressional, corporate, and industrial interests, not by listening to a few cranks in a small town where they might have made a mistake.

Still, the port manager's capitulation to the considerable concentration of wealth and power downstream is mainly bilge water. He's not advocating for local control, unless it's the kind of local control that earns the Corps and its constituents what they want. The Corps and the commercial interests represented by the port have thus far managed to maintain thorough control over management of the river's flow, even as the silt continues to pile up, even as local residents continue to bankroll the port's operation out of their taxes. Far from voicing the frustration of being a powerless minor bureaucrat, Doeringsfeld can wash his hands of the salmon issue because his job requires his loyalties to be aligned with interests outside the community, with the defendants in those court cases he claims are making the big decisions in the posh metropolises of the Coast. He might be right to say that Lewiston was sold down the river. But it wasn't any salmon lawyer who executed the deal.

~ ~ ~

Doeringsfeld and the Port of Lewiston, along side every other "port" in the Columbia Basin, belongs to the Pacific Northwest Waterways Association (PNWA). Its stated mission is to "collaborate with the U.S. Congress, federal agencies, and regional leaders on policies related to transportation, energy, trade, and environmental policy." Put more bluntly, the organization functions as the lobbying arm of the Corps. It includes corporations like CMH2Hill and Weyerhaeuser, every municipal and regional port that conducts business along the length of the Columbia, public and private utilities (Seattle Public Utilities is a member), the barge companies that operate on the river, and several prominent law firms that defend the interests of these concerns. Groups like the PNWA and the Northwest River Partners lobby for increased funding for the Corps, increased subsidies for the companies that rely on the work of the Corps, and salmon recovery, as long as such efforts don't interfere with the work of the river as a servant of the industrial economy. This is bad news for salmon, and it does not bode well for an equitable solution to Lewiston's silt problem. Each party is doing its level best to ensure that no outside influence threatens the monopoly this public-private mutualism maintains, whether the threat is genuine salmon recovery or a legitimate fix for Lewiston's levees. The capital the Snake River produces—hydroelectricity and a navigable waterway—gets exported downstream. Lewiston quite literally gets the dregs.

One aspect of this pecking order has become an issue for Thomas Dechert, who manages Lewiston's storm-water drainage system. I met Dechert the next afternoon for an extended tour of Lewiston's levees. As we crossed the blue bridge from the Washington side of the Snake over to Lewiston, Dechert recalled that in 2007, the Corps, along with city officials and a representative from the Boise office of the Federal Emergency Management Agency (FEMA) got together to inspect and discuss the levee issue. "The Corps maintains the condition of the levees themselves is good to excellent," said Dechert, "but the city is concerned about what will happen when the levees are overtopped." Like Doeringsfeld, Dechert is somewhat frustrated by the injunction against dredging. He notes with ambivalence that the City Council is not anywhere near endorsing a dam-breaching scenario. I relate to him Doeringsfeld's prognostication that no environmentalists will be available when the time comes to fill and stack sandbags. "That may be true, but in the meantime, where are we getting the sandbags? And where is the sand?" Dechert asked. Should a flood come to pass, Dechert pointed out, there is a disturbing dearth of emergency planning. During a multiagency levee tour in 2007, officials recalled that the Corps had once formulated a plan for emergency procedures, but whether the plan was on file in the Corps' local office or a hundred miles downstream at the Walla Walla regional headquarters was not known. The lack of a document from which to proceed, however, is hardly the most conspicuous oversight.

Dechert drove us through east Lewiston, past a residential neighborhood, then beyond a railroad switching yard where the city's water treatment plant was relocated after the dam was built. A half-mile downstream, a berm on one side and a bridge abutment on the other narrowed a gravel road. "Right here is where the Corps planned to separate east Lewiston from downtown in the event of a flood," pointed out Dechert, bringing the car to a stop. The plan, he told me, consists of using a bulldozer to plug the narrowest point along this access road, which happens to be where we're parked at the moment. "No one seems to think this would work," Dechert said, "but even if it did, where is the dirt and the heavy equipment?"

Dechert ticked off items from a punch list of other issues. The levees themselves have made the task of draining storm water more formidable. This trouble isn't hard to visualize. The levees keep river water outside the town but also keep rain water inside them. Most of downtown Lewiston now sits several feet below the surface of the water that flanks it on two sides. The Corps' solution was to create a series of ponds inside of the levees along the western (Snake River) shore of town, where water could be collected, filtered, and pumped through the levees and back out into the river. But the pumps that shunt water from the city's drains adjacent to the levees are past their useful life, with no plans to replace them anytime soon. The pumps are electric; no backup power source exists should the power fail during a flood. Emergency procedures, Dechert had learned recently, would involve operating a set of valves along the length of the levees by hand. No one could say for sure where these valves could be accessed, much less whether or not they would work. Inside the levees themselves, pipes are rusting, discharging an unpleasant, yellowish stream. The EPA wants something done about this. Upstream, a cement plant leasing its land from the Corps keeps contaminated wastewater in a pond adjacent to the river; should the substandard dike separating the pond from the river overtop or fail, millions of gallons of polluted water would head down the Snake.

The Corps has tried to make the case that it was not responsible for managing the storm water that drains through the levees, and that the city should be working solely with the EPA on the matter. So Dechert has been working overtime to solve the problem. He pointed out two places along the levees where the Corps installed devices that separate the oil that gathers on streets from rainwater draining from those surfaces. "The Corps says they don't manage storm water, but the infrastructure they've put in indicates otherwise," Dechert observed.

None of these problems would be insurmountable, attested Dechert, if the Corps was not so thoroughly unresponsive to requests by the City of Lewiston for help of any kind. "The only reason they're talking to us now is they're not getting the funding they need for the dredging for navigation," he explained, "so they thought they would come to the city to get them to put the pressure on legislators to get them the funding they want. But now that the city's become aware of some of the issues, there are some things we want from them, and they're not willing to talk to us at all about that."

The only things Dechert wanted were a viable emergency

plan in case of a flood and a clear delegation of responsibilities for managing the city's runoff. Instead, the Corps has told Dechert in the codified vernacular of hostile bureaucratic exchange to fuck off. In an e-mail correspondence between Walla Walla's deputy district engineer, Alan W. Feistner, and Dechert, which was copied to Idaho's congressional delegation as well as to Corps staff, Feistner's position on further talks was outlined in no uncertain terms:

> As I wrestle with your [Dechert's] request to talk more, I have come to the conclusion that my staff has adequately provided the legal and practical aspects of Corps' involvement for both PSMP [Programmatic Sediment Management Plan] and storm water. Therefore, for all your questions with regard to storm water, please work with the Environmental Protection Agency (EPA). The Corps does NOT regulate storm water. . . . In short, you know what we know. While I understand you do not agree with the facts we've provided, I don't know what further discussion would do other than take a lot more time and likely accomplish little.[5]

But as it turns out, the city did not know everything the Corps knew at the time. In a document filed away in the Corps' regional library, in a design memorandum, which delegates responsibilities before the blueprint for a project becomes a reality, specific lines dictate that the Corps *is* on the hook for storm-water management inside the levees as well as outside. The memorandum is prima facie evidence that the Corps, not the city, should be required to apply for a permit from the EPA and operate within the provisions of federal law, as any entity responsible for managing runoff must do. But unless a higher authority steps into the fracas, Lewiston may have to take whatever consolation it can get from these smaller rhetorical victories. As Dechert explained later to me in an e-mail, "The hang-up is that EPA does not want to come to legal blows with the Corps over the issue, and the city has been advised that we do not have deep enough pockets to come to legal blows with the Corps over it, so Feistner gets away with his distortion of the facts."

The rancorous flap over municipal runoff raises the question: why is the Corps so determined to stay out of the business of managing the city's storm water? After all, it's the Corps' levees that make draining the city more complicated. To an agency with a multibillion-dollar budget, the cost would be negligible. And the added responsibility would be in line with an agency that's always been adept at finding ways to justify ever more projects, repairs, improvements, and upgrades. Dechert had his own hunches on this matter, a point he saved for the last stop on our tour, an overlook of the place where the Snake and Clearwater join.

"The Corps seems to be totally unconcerned about whether or not downtown Lewiston will flood," assessed Dechert. "It's almost like flooding downtown Lewiston is part of their flood-management plan." We walked to the point overlooking the confluence. "When we did the tour with FEMA and the Corps, it was brought up that there's supposed to be seven feet of freeboard right here," recalled Dechert. We watched the water lapping on the shore not four feet below our shoes. "I've also seen this area from a boat when the Corps had it drawn down for inspection," he added. "We had to stay right out in the middle of the river, where the navigation channel is. The rest is silt." I ask Dechert if there's been any data gathered on what would happen in a ten-, twenty-five-, or hundred-year flood. "We can't get that data out of the Corps," said Dechert, "And the Corps says it can't give us that information until they get approval from the courts."

The darker possibility is that the Corps is withholding data that would indicate the flood risk is much more significant than they're publicly admitting. In any case, Dechert is pessimistic that the Corp possesses either the will or the ability to produce a plan for the sediment that will meet the requirements of federal law—the Clean Water Act (CWA) and the ESA. "They basically admitted to us that they're starting from zero, trying to come up with a model that will predict the sediment loads for the entire Snake and Clearwater basins," said Dechert, "and in the meantime, the Corps won't give us anything. Here's Lewiston, saying give us some idea of the risk, even if it's a wild-assed guess, so we can do some emergency planning."

The wind picked up, and from our vantage point overlooking the confluence, it appeared that rain would arrive in the Palouse that night as predicted. As we scurried back to the car, Dechert pointed out the abandoned cannery across the street from where we parked. "That complex is for sale," he said. "People come in here thinking they'll buy it, raze it, and maybe turn it into a nice little resort area, looking out over the water. But the minute they ask about the flood risk, forget it. It's dead."

Dechert drove us back across the blue bridge, and he and I parted ways shortly thereafter. Later I came back to that old cannery, ambling around its perimeter in the twilight. The whitewashed building sits downtown like a sad apparition, the ghost of promises unfulfilled. It was an easy spot to imagine filled with eager diners looking out over the water at the sunset. But there was nothing there. A streetlight buzzed. The wind whistled through rows of hollow diamonds in the chain-link fence that surrounded the place. A deserted waterfront block, where any urban planner worth his salt might locate the heart of the city, is probably not what the Corps and its boosters in commerce and industry had in mind fifty years ago. But this is what the dams have wrought: rather than creating wealth in Lewiston, they've become a hindrance to its realization.

~ ~ ~

Lest what I've described thus far should lead to any false impressions, I should point out two things: a good meal can be had in downtown Lewiston, and not everyone is stricken by suspicion or indifference to the plan to breach the lower Snake dams. On this point, Dustin Aherin didn't look like he would be the man to stump for freeing the Snake. By appearances anyway, he was not much different from any other thirty-something male in town. Hair close-cropped, wearing a sweatshirt and jeans, driving a big Dodge pickup, Aherin looked like he could be on his way to or from hunting camp or a construction job. The latter happened to be where he really was coming from. "I'm trying to finish building my house," he explained. Some of Aherin's preferences reflected his roots in small-town Lewiston, a trait he expressed when I asked where we should meet. When I first asked him where we should meet, Aherin capitulated to a strain of familial nepotism that seemed suitably podunk. "Well, my brother owns a restaurant in downtown Lewiston, so if we go anyplace else it might cause a family squabble," he told me half seriously. At his brother's place, however, it became immediately clear that Aherin was in some ways cut from a different cloth.

"Afternoon, Donna," he greeted the waitress. "I'll have that chicken sandwich."

"The bantam? Again?" Donna teased.

"Well, you know I'm not eating any of my brother's beef until he starts buying it from down the road."

"Fine," Donna sighed. "I'll keep my promise I made not to tell you anything bad about the chicken."

In addition to his preference for local cow, Aherin had, for better than a decade, been one of Lewiston's more outspoken proponents of breaching the lower Snake River dams. He was a river guide for nearly twenty summers. He also started a local concern called Citizens for Progress, who were working not only to free the river from the dams but the local economy from what Aherin saw as its narrow focus on barging wheat. "Nothing I've said or done seems to have pissed anyone off too badly," he said as our food arrived. "Well, that's not true," he immediately demurred. "People like Dave [Doeringsfeld] get mad when I write letters to the editor or editorials that point out how little the dams have helped our economy."

Aherin outlined the numbers on his side: all told, he calculated the federal government spent at least $250 million annually on the Snake dams alone, all to ship a roughly equal dollar value of soft white wheat, an export commodity that sells, with further subsidies, for about six dollars a bushel. "Farmers are business people, and what they've figured out is that the market is not there for all of them to grow the same variety of wheat," Aherin observed. "I have a friend up in Genesee, he's on about five thousand acres. He grows this variety of wheat they use in the artisan breads, baguettes, and that kind of thing. Well, last year he sold at about three times per bushel what they get for soft white. His wheat all goes to Spokane and then Seattle by truck and train. Portland doesn't have the facility to handle it."

These specialty varieties of wheat, Aherin argued, were part of a trend that was leaving Lewiston's port facility behind: "The Asian markets, especially as their economies expand, don't want one kind of wheat. A city in northern Japan likes a kind that makes their noodles soft; another in India prefers one with higher protein content. You can't just load up a barge with this stuff. Japan wants maybe an eighth of a barge load. You could put it in a container, but Lewiston has a really slow container operation." Because Lewiston had thus far refused to acknowledge these shifts in the global wheat markets, according to Aherin, other regional ports were stealing business out from under it. "Spokane's been getting together a proposal for a port authority up there, based on rail and the unit [high-volume] trains," he said. "When they pull that together, you'll see more people trucking their crop there. The problem is that the port isn't looking at all their options. I think there's this mentality there that to expand past the barges would be to admit defeat."

The long-term economic plan for Lewiston, maintained Aherin, should also be looking beyond just the port. "Recreation is now the third biggest segment of Idaho's economy," he explained. "When the fishing's good, you'll see fifteen guys here a night, all of them who've been out on the rivers with an outfitter, at three hundred dollars a day. You hear it all the time around here, that you can't build an economy around recreation. But look at a town like Missoula [Montana]. Not only do people figure out that these places are nice to visit, but they wind up staying there."

In Aherin's vision for Lewiston's future, the basis for creating a more diverse economy would involve renewing the city's ties to the river. He'd been collecting historical anecdotes about the river in the pre-dam era from the collective memory bank of Lewiston family and friends old enough to remember. "With the older folks, saving salmon can be kind of a tough sell," related Aherin. "My grandmother, she told me they ate so much salmon growing up that she got kind of sick of it. Salmon was what you ate when you didn't have enough money to go to the store and buy something else."

What seemed to resonate in people's memory was the city's former physical proximity to the river, a civic amenity most people Aherin had spoken with seemed to openly regret losing. "Donna over there," Aherin gestured in the direction of our waitress, "she remembers the big white sand beach that used to be on the Clarkston side, right beneath the blue bridge. On a summer day, there'd sometimes be a thousand people there. Other people don't want the dams torn out because they like to water-ski on the reservoir. Well I've got pictures of people waterskiing on the river below town before the dams went in."

To prevent such memories from deteriorating into nostalgia, Aherin acknowledged he'd be fighting the recalcitrance of the Corps as much as the status quo of his small town. We discussed the possibility that the Corps would get its way with dredging or raising the levees. "I would say that opposition to raising the levees around here is running better than ninety percent," he said. "I think people realize this just isn't a long-term solution. You're just raising your bathtub walls higher, which will just make a flood that much worse. This is my brother's restaurant, and you have to get on the roof of it to be at the same elevation as the river outside the levees."

Why then, I asked Aherin, was the Corps being so stingy about sharing data and dragging its feet on a sediment management plan no one thought would work?

"I've been in meetings where the Corps was present, and the subject of flooding risk came up," Aherin recalled. "I can't really follow their assessment of the silt buildup. So I asked if they could put things in terms of hydrology. Is there a peak flow where the city would be at risk? No answer. What it came down to was this: if it ever looked like Lewiston was going to flood, all three hundred Corps employees at the Walla Walla office would be put on buses and brought here to help fill up sandbags."

Undaunted by the Corps' ambivalence about what constituted an emergency and its reluctance to share information, Aherin saw the salmon's role in restoring the river as a kind of third rail on the track to getting the dams breached. Interest from outside the region was focused mostly on whether Snake River fish would be allowed to go extinct. Aherin doesn't discount the opinions of outsiders. To him, the survival of salmon ought to be a national concern.

"Those dams kill the fish," he said. "And we as a species and we as Americans made a decision to kill those fish. We still have an opportunity to reverse that decision. But it's coming down to maybe not enough time. I think people are beginning to realize we have the power to fix this. Nobody's going to get hurt. Everyone's going to come out whole. So let's fix it."

The belief that drove Aherin's activism, he promised, wouldn't be waning in the near or distant future. "I'm not going away," he emphasized. "I'm building a house here, and as long as the dams are here and I'm here, I'll be fighting for some better alternative. And that'll be true even if the salmon go extinct. The barges have never worked. Their heyday is long gone."

"To fix," it occurred to me after lunch, driving across the blue bridge and west out of Lewiston, is a verb with more than one meaning. For the time being, it seemed a fair conclusion that the Corps and the interests it represents are operating under the terms implied by the sense of the word *fix* that denotes collusion. Aherin keeps the faith. He can talk about the possibility of a salmon summit, when by judges' orders or by consensus, stakeholders in the great salmon debate might get together and hammer out some kind of agreement that would satisfy at least some of the wants and needs of each party.

But the idea of negotiation ought to imply a bargaining table on the level, around which each side brings something they are willing to part with and something they would dearly like to keep. And though political advantages can change at the drop of a hat, terms for fair negotiation don't seem imminent. The Corps is holding all the cards in a game it sees as already fixed.

Down at the river, the Columbia Queen, a replica of a Twain-era Mississippi gambling boat eased its way to the dock in Clarkston. The sternwheeler makes a multiday round-trip cruise from just east of Portland to Lewiston and back. From Highway 12, I glanced at what appeared to be a few disembarked passengers, who must have just churned past the buried treasures at the erstwhile Marmes Ranch at the mouth of the Palouse River. A nostalgic Southern-style cruise might be just the ticket to commemorate the dams and levies of the Snake River. Prodigious snow and rain in winter and spring in Idaho's mountains pose the flood risks this region faces each spring, much as the Gulf Coast endures hurricane season every year. Hurricane Katrina exposed the Corps' incompetence with levees there. Now Lewiston seems like a good bet to become the next New Orleans.

8

THE FIFTH H

*Water in the capitalistic state has no intrinsic
value, no integrity that must be respected. . . .
It has now become a commodity that is bought and sold
and used to make other commodities that can be
bought and sold and carried to the marketplace.
It is, in other words, purely and abstractly a
commercial instrument. All mystery disappears from its
depths, all gods depart, all contemplation of its flow ceases.
It becomes so many "acre-feet" banked in an account,
so many "kilowatt-hours" of generating capacity to be
spent, so many bales of cotton or carloads of oranges to
be traded around the globe. And in that new language
of market calculation lies an assertion of ultimate
power over nature—of a domination that is
absolute, total, and free from all restraint.*

—DONALD WORSTER,
Rivers of Empire

In the tumultuous summer of 1968, Ed Chaney was witness to
a slaughter. Three hundred thousand dead salmon and steelhead were rotting downstream of the just-completed John Day
Dam, thirty miles upriver from The Dalles, Oregon, some of
the carcasses laid in so thick they looked to him like someone
had stacked them like windrows on the beaches. Two years earlier, the Oregon Fish Commission (later the Oregon Fish and
Wildlife Commission) had hired Chaney, a rookie fisheries specialist, to work out of its Portland office, where one of his tasks
was to report on fish passage issues on the Columbia, including
work at the new dams.

"I was chewed up and spit out," Chaney told me. "They had designed that dam to waste as little water as possible on anything other than power production. The entrance to the fish ladder was not accessible for fish. So the Corps agreed to install pumps to move about a foot of water so the fish could find the entrance. Well, the pumps were ordered but delivery was delayed. But the vice president [Hubert Humphrey] was scheduled to come out and dedicate the dam. The Corps sent us a letter that said, in effect, we're closing the dam anyway."

When the Corps did so, without the requisite pumps, or even turbines in some of the bays at the powerhouse, the massive waterfall created below the dam caused a phenomenon known as nitrogen supersaturation, the piscine equivalent of what deep-sea divers call the bends. The Corps hoped mightily that no one would notice. "At first, the Corps of Engineers wouldn't even allow state biologists onto the dam. Here I am, this kid from Missouri, wondering what's going on here?" recollected Chaney. The Corps, he soon realized, was engineering a cover-up. "The *Oregonian* came out and interviewed me, and I told them what I saw," he went on. "It was a big story, and the next day on the front page, there's the Corps' district engineer bad-mouthing me, saying it didn't happen this way, I was speaking out of turn, I didn't know what I was talking about." Chaney attempted to write a press release to clear his name of the accusations the Corps leveled at him, but his boss heavily edited several versions, telling Chaney he wanted to "preserve our good working relationship to the Corps."

Frustrated by the censorship, Chaney surreptitiously took a camera and photographed some of the salmon carnage, then took two rolls of undeveloped film to the offices of the *Oregonian*. "I told their reporter: 'I'll give you these on one condition. You didn't get them from me, and you don't know who I am or where they came from,'" he explained. "They agreed. Well, I left, thinking, I've done all I can do. The next morning, I see another front-page article on the disaster. It says go to such and such a page for photos. So I go there, and there's a full page of photos and a little box in the center of the page that says, 'Photos courtesy of Ed Chaney.'" The Associated Press picked up the photos, which ran nationwide.

After a brief visit to the Oregon governor's office, it became clear Chaney would have to move on. He became a fisheries consultant, working with tribes and conservationists to maintain and reestablish lost salmon runs, eventually focusing his efforts in Idaho, Eastern Washington, and Eastern Oregon. But in his most prominent role, he's become something of a bio-political activist, speaking out against the Corps and the BPA colorfully enough that *60 Minutes* once came to interview him for a segment on the salmon crisis. More recently, he brokered a deal to put salmon back into the Umatilla River, which involved getting water back into a river that had been chronically sucked dry by farmers for nearly a century. "They fish for them right in downtown Pendleton sometimes now," Chaney boasted. "One spot they call the Taco Bell Hole, which is where you park to get to the river."

His acerbic wit is piqued by a lunch-hour discussion of a long-running BPA public relations campaign. Saving salmon, according to this plan, comes down to the "4 H's": habitat, hydro, hatcheries, and harvest. This strategy has been promoted with the same gleeful boosterism found in the bucolic projects of the national farm organization with the same 4-H acronym. Three of the four H's, according to the BPA's scheme, should receive the benefit of immediate reform. Habitat lost to logging, mining, agriculture, and development should be repaired. Hatcheries should operate on the frontiers of conservation biology but not sacrifice industrial efficiency. Harvesting should be cut back. Hydropower production, by contrast, should be maintained at current levels at virtually any cost.

It's the hydrosystem's version of parental tough love for its troubled dependents: plenty of love for the folks, tough luck for the kids. What hydrosystem defenders like to call their "spread the risk" philosophy is fraught with potential pitfalls for fish but contains quite tangible benefits for dams. Under the guise of this holistic-sounding approach to restoring salmon, solutions to the problems that dams alone have created for salmon can be neatly deflected onto tributaries, estuaries, the ocean, fishermen, global warming, or the host of salmon predators that hunt the river.

Chaney contends there is a fifth H.

Being a man of letters and not a scientist, I was intrigued by this idea, recollecting that some Indians who fished the falls at Celilo assigned a religious significance to the number five. Could the fifth H be a New-Age, anadromous Third Way, another iteration of the tao of physics, a corroboration of the impending harmonic convergence between science and ancient wisdom? The answer was revealed to me in church—the church that had been converted into a deli where we were having lunch. "The fifth H," Chaney explained to me, a clergy-esque trace of genteel Missouri accent still evident between bites of sandwich, "is horseshit, since the BPA has always far understated the impact that dams have on salmon."

Chaney may be angry as a cut snake, but he's accurate as a scientist. According to recent numbers, gleaned from an interagency, decade-long endeavor dubbed the Comparative Survival Study, a very conservative estimate for current overall salmon mortality due to the hydrosystem in any given year would be half the population. A 1994 Ninth Circuit court decision assigned the blame attributable to dams for that overall decline at 80 percent. Chaney pointed out that all other stakeholders in the Columbia—from grubby gillnetters working the estuary in Cathlamet, Washington, to the spey-rod swinging yuppies in a riffle on the Clearwater, in Idaho—were left to fight among themselves for their share of the remainder. And that amid this squabbling, Chaney said, too little attention has been paid to a last-ditch salmon barging experiment the federal family has since upheld for three decades as a viable restoration plan.

In the mid-1970s, some horrible drought years beset the Pacific Northwest, steepening the alarming descent of Snake

River salmon stock toward extinction just as the last four federal dams on that river were completed. The decision was made to gather juvenile salmon at the dams and barge or truck them downstream below the tailrace of Bonneville Dam. So it was that traveling Interstate 84 back then, one could witness a sign of ecological turmoil as surreal as the image of the Cuyahoga River on fire. Wheat, timber, and other commodities were being hauled up the river to the mountains of Idaho, while Idaho salmon swam in tanker trucks going the other way on the road. What began as a salvage operation, a dire action to prevent catastrophic mortality for fish, has morphed into the preferred plan for getting migrating salmon to the ocean. The day and a half ride on one of the Corps' smolt barges, from the Snake River to the lower Columbia, the Corps claims, boosts juvenile survival to near 100 percent. Survival is measured by the number of fish that survive the boat ride. This is like claiming a cure for cancer by pointing out all the terminally ill patients who survived the van ride from hospice to the chemo ward. Survival of barged juvenile fish to returning adult is not much different than for fish that manage to stay in the river. Over the long haul, neither group has survived well enough to reverse their dwindling populations. "Barging is a clear sign that the river is too lethal for fish to swim in," Chaney said. "Some scientists say they can calculate the benefits of barging. But the more deadly in-river conditions, the more benefit barging appears to have. The feds haven't been barging endangered Idaho fish for thirty years. They've been barging for thirty years and it landed the fish on the endangered species list. Usually, results count. Not in this case."

Dam advocates' rationale for salmon demise leaves Chaney cold. "You hear these people say it all the time," said Chaney. "'We just didn't know any better.' That has simply not been the case." Chaney realized after his run-in with the Corps at John Day that nothing would be accomplished on behalf of salmon unless the forces maintaining a stranglehold on the river were reformed. He admitted he has largely struck out in this endeavor. Years of scientific scrutiny have improved the understanding of the biology that drives salmonids from rivers to the ocean and back again. But the politics, in his view, are not much different than they were in the summer of 1968. The biological science of saving salmon has become a tool in the political science of not saving them. "They furnish all their statistics and results of their studies," Chaney asserted, "and then we argue about the numbers. But the argument ought not to be about that. The argument should be over *assumptions*.

"Look at it this way," he continued, grabbing my keys. "I steal your car. And I start a taxi service with it. You call the cops thinking they'll have me arrested. But instead, they start studying the way my new business runs. So then they tell you: 'Sorry, you can't have your car back. Your car's turning a pretty good profit for this fella here who stole it. Guess you'll have to wait until he's done with it, then maybe we can talk about getting it back to you.' The hydropower interests are basically saying: 'This is our river. We stole it fair and square. We're not giving it back.'"

The grand-theft hydro analogy isn't all hyperbole. In 1995, the prestigious National Research Council (NRC) issued a report criticizing the hydrosystem for hogging all the water. Its authors dryly concluded:

> The largest annual cost for salmon restoration in the Columbia River is water. An implicit assumption in the design of the system of dams was that there would always be water for fish. Water, however, is not available in the right places, in adequate quantities, or at the right times to meet the needs of migrating salmon. Fish have no rights to Columbia River water. To implement restoration goals, water must be purchased from those who control it.

The report refers not to questionable science methods but a creative accounting practice. Those who shill for maintaining the status quo on the river are eager to repeat the staggering sum of money the BPA has spent saving salmon. Soon the bill will top $10 billion over the past thirty years, and annual expenditures have ballooned to more than $600 million. Yet according to the NPCC, more than half of this outlay goes to charging electricity ratepayers and federal taxpayers for allowing salmon to pass up and down the river. The cost for water that flows oceanward without first spinning a turbine, floating a barge through the dam's locks, or irrigating farmland is charged to the Fish and Wildlife Program, created to recover salmon on the Columbia and Snake rivers, and ultimately to ratepayers, as "foregone revenue" of which the dams' beneficiaries are deprived. Put yet another way, the BPA charges money for the river to run as a semblance of its former free-flowing self. According to this logic, every raindrop, snowflake, glacial drip, creek, brook, spring, seep, or tributary in the 258,000-square-mile watershed not claimed by a landowner with a prior water right belongs to the dams. To continue with Chaney's grand-theft hydro analogy, this is a practice akin to that hypothetical cab driver charging the city in which he operates the going fare for the hours the car is parked.

Conjuring up the scheme to validate the accounting has not been easy. But the consortium of federal agencies that manage the hydrosystem, including the Corps, has managed public perception as well as any modern military campaign.

Barging fish and charging citizens for the water fish use if they don't get on the boat are two of the three prongs in the feds' preferred approach to salmon recovery. The third is an alarming dependence on hatchery production.

Hatchery-raised salmon have some of the same problems as people who spend too much idle time in artificial environments. Physical fitness quickly becomes suspect. As David Montgomery writes in *King of Fish*,

Releasing hatchery fish into a stream is like dropping suburban teenagers into the middle of the Congo and asking them to walk out of the jungle to the coast. Few will make it. The hatchery fish that do make it back may well be suited for life in the marine environment, but the hidden price of reliance on hatchery fish is that resilience to disturbances, environmental change, and natural hazards in the equally crucial river environment may be bred out of a population.

Problems with breeding practices in hatcheries are unfortunately analogous to those of the fundamentalist Mormon practice of polygamy. Free agency in sexual selection is completely absent. Telltale signs of a family tree with a dearth of branches, far too narrow for its height—including missing fins, malformed spines, and other ill effects of inbreeding—occur too frequently.

A lack of wile and a kissing cousin syndrome are only a couple of the bad influences hatchery fish bring. Smolts are allowed to mature in the shelter of hatcheries longer than they would stay in protective stream reaches in the wild. Their generally larger size and sheer numbers give them an advantage in competing with smaller, less populous wild fish for food and habitat. Like hordes of conquistadors unwittingly armed with a resistance to smallpox, hatchery salmon also have much greater odds of wiping out native fish by introducing disease. In hatchery facilities, fish are inoculated against illness; for practical reasons, wild fish are not.

Nonetheless, the most notable element in the response to diminishing salmon numbers has been to keep adding more hatcheries. In 1950, there were 28.9 million hatchery smolts released into the Columbia. By 1968, there were 92.7 million. Throughout the 1990s, more than 150 million a year were produced. In 2005, there were 134 million cut loose from more than two hundred hatchery facilities around the Columbia Basin. Yet none of the historic productivity, much less genetic diversity, lost to dam construction has been won back. Thus, the Snake River Basin, which once produced in excess of 2 million wild chinook annually, now produces less than 1 percent of that historical figure each year. Based on slightly improved returns since 2005, when a federal court ordered the dams to release more water, the feds have touted these salmon as recovering. In fact, the best that can be said is that they're not headed for extinction as quickly as they would have been in years past, when returns were counted in the hundreds.

As wild fish continue their decline, it's not as if defenders of the hydrosystem will be able to say they weren't warned. The NRC declared that the Columbia River hatchery system has operated out of "deep ignorance," with utter disregard for even the most basic scientific principles. It recommended a No Child Left Behind approach to reform, with failing hatcheries to be upbraided as harshly as failing schools:

> Hatchery operations should be changed in accordance with the goal of rehabilitation and the ecological and genetic ideas that inform that goal. . . . Dismantle hatcheries where they interfere with a comprehensive rehabilitation strategy. Hatcheries should be rigorously audited for their ability to prevent demographic genetic fish health, behavioral, physiological and ecological problems. Any hatchery that mines broodstock from wild spawning populations should be a candidate for immediate closure or conversion to research.

Rehabilitation, not mere mitigation, wrote the NRC, should become the goal.

But like a determined addict, the federal family has clung to its stalwart belief that rehab is for quitters. Three-quarters of all Columbia and Snake rivers salmon are now hatchery fish, a level of dependence that many biologists see as an impediment as large as the dams to meaningful recovery. But there's no easy way to get the hatchery monkey off the river's back. Going cold turkey would reveal the full extent of the river's transformation from a wild ecosystem to an indentured servant of the industrial economy. Take away those genetically inferior fish and a surly hodgepodge of angry, hungry, out-of-work commercial, sport, and tribal fishermen might combine forces with their recently laid-off hatchery comrades and begin organizing in dangerous or unpredictable ways. Hatcheries have become the opiate of many fishermen.

Chaney believes that the federal refusal to try anything new has dropped the dam-removal argument in the laps of salmon advocates. "I've watched the idea of dam removal gain traction as the feds stuck to their game plan," Chaney said. "Truth be known, there was a time when this could have been solved, and dam removal might have never been an issue."

By simply providing a dedicated, dependable flow for salmon —as Chaney recalled fisheries managers were asking of the feds more than thirty years ago—the fix might have been in. But water for salmon would mean returning a portion of the river that was "stolen fair and square." Instead, the feds came to depend on more spin.

In 1999, the Corps completed a four-year, $20 million study that bucked the trend of economists and scientists whose calculations seemed to make the prospect of large-dam removal on the Snake feasible. The Corps concluded that breaching would have a net economic cost of $246 million per year. This was the new math. Forty years prior, in its own cost-benefit analysis of the dams, the Corps had reported that taxpayers would see only fifteen cents in benefits for every dollar spent on lower Snake River dam construction. From other federal agencies, and even within its own camp, the Corps quickly came under fire for cooking its books. The EPA reviewed the Corps' dam breaching study and faulted it for "[failure] to interpret or resolve contradictory data and conclusions . . . selective use of data . . . faulty or misleading interpretation of results . . . failure to present important information or assumptions . . . and unsupported conclusions." The EPA excoriated the Corps for excluding from its analysis any reckoning of the cost of complying with the Clean Water Act. Short of removing the dams, complying with this law, it was estimated by the EPA, would cost somewhere between $460 million and $900 million a year. All four reservoirs routinely violate water quality standards because of high temperatures. The EPA also noted that short of dam breaching, there was really no cost-effective alternative to meeting federal water quality standards.

Other economists who vetted the study, most notably University of Oregon economist Ed Whitelaw, reckoned that most of the costs the Corps used—from replacing the lost electricity generation to restoring riparian habitat along the re-exposed banks of the river—were wildly inflated. To calculate the economic hit from potential job losses, for example, the Corps assumed anyone laid off by the disappearance of the dam's infrastructure would never find another job. Moreover, it assumed the penalties of perpetual joblessness would extend beyond any reasonable life expectancy. By the Corps' numbers, un-damming the Snake would damn displaced workers to an eternal hell of unemployment, for which seemingly immortal citizens would pay penance.

The Corps even managed to ignore its own numbers. Among the most glaring omissions was the calculation of what economists call "passive values." These are parameters that put a dollar amount on intangibles such as the worth of sites of cultural, religious, and biological heritage. Detractors see the practice as a blasphemous attempt to place a price tag on places like the Vatican or Wrigley Field or Yosemite National Park. An appraisal implies the goods in question are for sale, when, to those who hold these places dear, such sacred locales are better esteemed priceless. Nonetheless, these kinds of valuations are an accepted feature of modern economic studies. The Corps' own recreation planner in the Walla Walla District at the time, 1999, a man named Phil Benge, worked with Dr. John Loomis, an economist at Colorado State University, to estimate the passive values that the free-flowing Snake, along with restored salmon and steelhead populations, would produce. Their joint study boosted credibility for dam removal: they concluded benefits would range from $142 million to $508 million. Recreation benefits alone ranged from $70 million to $416 million. The Corps' top brass, at the behest of Senator Slade Gorton (Washington), heir to an East Coast fortune gained from peddling frozen fish sticks,[1] who then sat on the powerful Committee on Appropriations, insisted these values not be included in the Corps' study. Thousands of years of evidence from around the Pacific Rim to the contrary, Gorton was leaning on the Corps to say that the prospect of restored salmon runs was worth nothing culturally. The Corps eventually settled on a "middle value" on the matter of $82 million.

According to a series of articles in the *Washington Post*, both Benge and Loomis believed that Corps officials, under pressure from Gorton, manipulated, then misrepresented their results. "Gorton didn't want us to find out anything that might hurt his cause, and the generals didn't want to say no to him," Loomis summed up in the *Post*. "I guess they were afraid he'd cut their budget. It was a classic case of best professional practices saying one thing, and our fearless military leaders caving to politicians and doing something else."

Deep doubts linger on the ability of the Corps to render any analysis honestly. A Pentagon investigation found "widespread bias among the Corps employees interviewed" and concluded that most of the Corps' efforts were bent toward expanding its budget and the scope of its mission. An eagerness to please corporate patrons and congressional sponsors, the Pentagon found, had helped "create an atmosphere where objectivity in its analysis was placed in jeopardy." In an interview in that same report, a retired lead economist for the Corps called his own agency's studies "corrupt."

The Corps' record on weighing public opinion would certainly bear out this opinion. When sixty thousand Sierra Club members, inspired by an essay by the nation's leading literary icthyophile, David Duncan, took the trouble to sign and pay postage for a card calling for a serious look at Snake River dam removal, the Corps, noting the uniformity of the card, counted all sixty thousand responses as one person.[2]

To Chaney, the revelation that the Corps cheated the science on dam removal was no surprise. "They designed these four dams in direct defiance of the intent of Congress, which approved the dams only if the salmon runs would be maintained. Instead, they built them with no way for juvenile fish to get out to sea," Chaney said. The legislation authorizing McNary Dam and the four lower Snake River dams states that "adequate provision shall be made for the protection of anadromous fishes by affording them free access to their natural spawning grounds." For anadromous fish, that has to be a free-access round-trip. The dams as originally designed were illegal.

With Chaney's help, after the last of the four lower Snake dams was finished in 1975, an effort began in earnest to drag the Corps kicking and screaming into providing juvenile bypass systems that would be integrated into the structure of the dams. States, Indian tribes, and commercial- and sport-fishing interests insisted the Corps make right what had been built wrong. While not entirely pleased with the prospect, the Corps nonetheless went along with plans to retrofit the dams.

A fish bypass system gives young salmon three ways to get around a dam. One is through the aforementioned retrofitted bypass system. Fish must commit an unnatural act to use this escape route. Usually, salmonids prefer to stay near the surface of the water, no more than twenty feet deep. To find the bypass portal, salmon must dive up to eighty feet down into the forebay (upriver side) of a dam. At this depth lies the intake to the bypass system, which steers smolts clear of the more turbulent portion of the river's current diverted to spin hydroelectric turbines. Due to the depth and velocity of the pipe, the pressure on the fish is disorienting. But with a few detours, the intake will lead them safely across the portion of the dam where the turbines churn water to make electricity. With a short break at a fish-handling facility inside the dam, where specimens are anesthetized, weighed, measured, and stapled with a rice grain–sized transponder that helps track them like grocery store inventory, the fish are then ushered to the tailrace of the dam. They will either be let free to swim to the next dam, where the collection and bypass facility process will begin again, or put on a boat to save them the trouble.

If fish can't find the bypass intake, they might take their chances with a plunge over the dam's spillway. The fall is also disorienting, and the nitrogen supersaturation that Chaney documented can be fatal if spill is not calibrated to prevent the phenomenon from occurring. In recent years, a massive contraption dubbed a spillway weir has made the route over the dam more attractive for outgoing smolts. "Flip lips," protrusions in the face of the spillway that add enough texture to slow the velocity of falling water, also help prevent nitrogen problems.

The last route is through the powerhouse. This doesn't happen much anymore, since the turbine intakes for the powerhouse are screened to prevent it. But young salmon, pressed by water pressure against the screens, can be de-scaled. None of these accommodations were available when Chaney began working on behalf of parties who sought them. While not ideal, bypass systems and spillway modifications at dams, when coupled with dependable flow that mimics the behavior of a natural river, show some promise. All the Snake River dams now have bypass and collection facilities. If the dams come out in time to save Snake River stocks, it might be said that these measures kept the fish from going extinct. As Chaney points out, however, so far, all the modifications have done to date is buy dangerously thin fish populations a little more time. "All these add-ons could not overcome the basic design flaw. The Corps' design made no provision whatsoever for juvenile fish to migrate downstream. Tacking on flip lips, turbine intake screens, collection and transportation facilities, all that was like pasting butterfly wings on pyramids in hopes of making them fly," he said.

Even if political action could retrofit pyramids to make them fly, the Corps would likely find a way to clip the wings. In the late 1980s, as Snake River salmon plunged toward extinction and politicians were scrambling to avoid listings under the Endangered Species Act (ESA), two United States senators, James McClure (Idaho) and Mark Hatfield (Oregon), managed to pry

$8 million out of the federal budget to dedicate toward improving and increasing the quality and quantity of bypass systems on the Snake dams. Knowing about how the Corps preferred to operate, the senators made sure the language appropriating the funds came with specific instructions on how to spend the earmark. Crude bypass facilities had been constructed at Lower Granite, Little Goose, and Lower Monumental dams on the Snake. The senators wanted these facilities improved and a fourth built at the last dead end for juvenile fish, Ice Harbor Dam. But once again, the Corps defied congressional intent. Major General H. J. Hatch, the Corps' director of civil works, figured he could get by spending only half the money on bypass systems. The other half went to buying two more fish barges and research into the problem the Corps had created by not providing a thruway for ocean-ward salmon in the first place. The Corps claimed the wishes of Congress were not legally binding, an idea that the Reagan-era Office of Budget and Management saw as being in line with its thinking on the matter. As usual, the Corps got its way. The Republican senators, neither noted for his dedication to environmental matters, were left fuming.

About this time, Chancy realized the agencies that owned and operated the dams, as well as marketed and sold the electricity from them, had no intention of meekly complying with the host of federal laws and treaties that mandated salmon runs not only be prevented from going extinct but be recovered to the fullest extent possible. "In '73, the Endangered Species Act passed," Chaney recalled. "In the next few years, the Clean Water Act came through. Then in 1980, the Power Act passed." It was this last law that mandated that salmon recovery and power production in the Columbia Basin would thenceforth be considered on equal footing. Additionally, amid a national resurgence of Indian legal rights and cultural recognition, it became clear, based on a pair of key federal court rulings, that the river would have to bear enough salmon to satisfy conditions spelled out in treaties with the river tribes in the nineteenth century. An international salmon treaty with Canada was also inked. With

American territory squeezing the Canadians from the north and the south, American dams and fishermen were taking a disproportionate share of Pacific salmon. It was agreed that the United States should provide one fish from its Lower 48 territorial waters for every Canadian salmon that escaped to Alaskan water. The pressure on the hydrosystem to let the river produce some fish again was mounting. "So while all this was happening," Chaney recalled, "a guy from Idaho who served on the Pacific States Marine Fisheries Commission, Herman McDevitt, started threatening to sue under the Endangered Species Act to get protection for Idaho salmon. Of course, the feds and all the politicians were beside themselves, saying, 'Don't do it!' Partially in response to that threat came the deal for salmon in the Power Act."

Chaney was a player in crafting the portions of the law that were included to protect salmon. The law called for the creation of a power planning council (now the NPCC) that would make recommendations to agencies operating the dams. The advice of the council would help secure a reliable supply of electricity as well as take on the ambitious goal of doubling the basin's salmon returns from their 1980 level of around 2.5 million salmon a year to 5 million. Instead, the annual returns have been more than cut in half again, to just over a million salmon a year. Capitulating time and again to the whims of utilities and other industry players that enjoy the subsidy of the cheapest power in the nation, by 1990 it was clear that federal agencies had failed miserably at the ambitious restoration goal laid out in the Northwest Power Act (the NPA). Chaney claimed the NPCC created by the NPA shoulders much of the blame. "There are good people associated with the Council," Chaney asserted, "but there are also people there who work directly against any kind of change to the status quo, against salmon recovery, and against the law as stated in the Power Act."

Bitter changes to the once comparatively sweeter flavor of regional and national politics have doomed the NPCC in its mission to recover salmon, said Chaney. The four states directly affected by policy decisions in the basin—Idaho, Montana, Oregon, and Washington—are each represented by two council members appointed by the governors of their respective states. "Back when they started, the makeup of the voting members of the council, the ones who ultimately decide which policy recommendations to make, was vastly different. Oregon and Idaho openly favored salmon recovery, Washington was neutral, and all Montana cared was that their reservoirs weren't emptied every year to keep the other states in power and fish. Now every state but Oregon is opposed to half of the mission—the fish-and-wildlife half—the council was created to make happen. Funny how when the dams were built, there was no consideration of delegation of power to the states, but when it came to undoing the damage done, all of a sudden this was a pressing concern."

Seeing the handwriting on the wall as a conservative backlash swept the midterm elections in 1994, Chaney sued the NPCC in the Ninth Circuit court that year—and won. The court directed the council, in effect, to ignore the constant pressure applied to them, by agencies and their industry constituents, to consider cost as the limiting factor in following their federal mandate to write a plan that would actually recover fish. "Their reaction to this ruling," recalled Chaney, "was basically that they were pissed they lost in court. As far as salmon are concerned, nearly as I can tell they still have no plans to change anything about the way they conduct their business."

With no obvious changes to the status quo, Chaney feels he has no alternative but to keep fighting. "Rather than a legally sound plan based in science that will bring salmon back," he said, "the whole bar has been lowered to merely keeping fish from going extinct. The other laws and treaties, which set a higher standard, are just being ignored."

Worse, the lower standards made available by the ambiguities written into the ESA are not low enough for the federal family. None of the technological innovations in salmon science or dam engineering work without adequate water. How the hydrosystem operators have managed to keep all the water for themselves, in spite of a considerable volume of science that suggests they shouldn't, is yet another story.

Though the legal background is arcane, the scientific data complex, the alphabet soup of agency acronyms confounding, and the dollar amounts staggering, the formula is familiar and quite simple: follow the money. If repeating this mantra from Watergate's Deep Throat seems too melodramatic, the allusion still works as a salacious reference to the durable skin flick. A good place to begin unraveling this particular deception is with a certain disgraced Idaho senator with an alleged predilection for dalliances in airport bathrooms.

LIES, DAM LIES, AND STATISTICS:
The Science of Saving Big Hydro

We also know that the more people there are
on the government payroll, the more likely it is that
someone will be encouraged to take a bribe.

——BARACK OBAMA, August 28, 2006, campaign speech

It's not an illegal bribe, but it's a bribe.

——MICHAEL C. BLUMM, salmon policy expert and Lewis and Clark
Law School professor, commenting on the $43 million check the
State of Washington took in late 2009 from the BPA for joining them as
defendants in a lawsuit over dam operations on the Columbia River

Eighteen months before former United States senator Larry
Craig (Idaho) was squirming uncomfortably in the national
spotlight for his wide stance in a men's room at the Minneapolis
airport, he was guilty of a legislative sleight of hand on the floor
of the Senate. In December 2005, the disgraced senator, a top
recipient of money from the hydropower industry, inserted his
proverbial quill into a conference committee report for a larger
spending bill, effectively zeroing out the entire budget for a
small entity known as the Fish Passage Center, not only for
the fiscal year 2006 but for all future years. The center's only
mission for a quarter century had been to provide independent
scientific analysis of the condition of Columbia and Snake rivers
salmon runs.

The inception of the Fish Passage Center occurred in 1984, four years after Congress passed the Northwest Power Act (NPA). It was born of a need identified by states and tribes, who were leery of the hydropower-friendly science being pushed by the BPA even then, for an independent source of scientific information within the basin. The center and its researchers were charged with the duty of answering any salmon query, from any party, any time, and to make its reports open to all. Since then, the center has consistently been reviewed as competent, efficient, and credible. State, federal, and tribal wildlife agencies have come to rely on the Fish Passage Center to inform their often controversial and always high-stakes decision making.

Scientists at the Fish Passage Center have not backed down from publishing data that shows that the salmon restoration measures the BPA and the rest of the federal family prefers are not working well enough. Hatchery fish hauled downstream on barges have never returned as adults in significant enough numbers to fulfill the legal mandate to restore the runs. These assessments, in the eyes of the BPA, encourage the consideration of more drastic alternatives.

Short of tearing out dams, one such alternative is known, with a nod to Orwellian Newspeak, as spill. "Spilling" water over a dam is not, as the word suggests, an accident. Water routed over a dam rather than through turbines is "spilled," and to those dedicated to the uncompromising production of electricity, it becomes as useless as a puddle of milk on the kitchen floor. But where salmon are concerned, good evidence gathered and disseminated by the Fish Passage Center suggests spill is the next best thing to a free-flowing river. Since ordered in 2005 to spill by a federal court, dam operators have provided enough water that survival rates for out-migrating juvenile salmon have markedly improved. Sockeye salmon made it to Redfish Lake in Idaho in numbers not seen since before the dams were built. The Snake saw post–dam-era record returns for both steelhead and fall chinook. The BPA and its preferred customers, however, have always fought tooth and nail against any measure that compromises their ability to turn falling water into money. The return of sockeye to their wilderness origins, the BPA claimed without the accompanying data, was the result of improved hatchery procedures and ocean conditions. Lack of evidence, in this instance, did not imply a lack of procedural options.

After federal judge James Redden ordered the very kind of spill the BPA was loathe to commit—mandatory consistent flows not subject to the BPA's fiduciary whims—the hydropowered wheels that spin truth and stifle scientific research went into overdrive. The Fish Passage Center had been asked to make a preliminary check on the efficacy of spill. Even the earliest results looked positive for salmon. Hydro-power's solution to this problem was simply to kill the messenger, via Craig's position on the Senate Committee on Appropriations.

In providing the rationale for this move, Craig cleverly cut up quotes from an independent scientific advisory panel that evaluated the Fish Passage Center in 2003 to make it appear, as has been the claim of the power industry, that the center operates as an ally for the save-the-salmon cause. "Many questions have arisen regarding the reliability of the technical data [the Fish Passage Center publishes]," Craig cited from the review. But the report actually stated the center's work was of "high technical quality" and that the questions regarding the reliability of the center's work were without merit. One author of that report called Craig's remarks "a misrepresentation." The damage, however, was already done. "Data cloaked in advocacy create confusion," said Craig. "False science leads people to false choices."

While what causes some people to speak falsely remains a topic of considerable ambiguity with the former Idaho senator, the BPA was happy to proceed with business as usual without the prospect of the Fish Passage Center's data contradicting its claims.

Tribal, environmental, and fishing groups sued the BPA over the move to rid itself of the Fish Passage Center, and a panel of Ninth Circuit court judges found in their favor. Dismantling the center, the three-judge panel found, was at least as illegal as soliciting a bathroom quickie from an undercover cop. The court found that Craig's stealthy efforts to ream the center did not carry the force of law. The BPA was admonished by the court for its "slavish adherence to a sentence in a legislative committee report" and instructed to restore funding to the agency, effective immediately.

The lengths to which defenders of the hydropower status quo were willing to go to silence a few scientists was out of proportion to the political power the Fish Passage Center can wield. With a staff of a dozen scientists and an annual budget of $1.3 million, it has no jurisdiction over policy-making and, if history is any indication, very little influence over decisions on how the river is ultimately managed. What galled the BPA and its defenders was the evidence the center offered against the mainstay of the hydro-power industry's mythology: salmon don't need a river.

Michele DeHart is the manager of the Fish Passage Center, a job she's held since the center was created. Over the years, she's cultivated her own response to the "4 H's" that she calls the "4 D's": delusion, denial, dithering, and decision. The mnemonic pertains not so much to salmon as to making tough choices, some of which happen to be informed by science. "With Columbia River salmon, we are in the dithering stage," assessed DeHart. "If the salmon crisis was your own little personal crisis, something that was happening in your life, you'd probably be at the point where you'd want to be doing something about it." The state of salmon on the Columbia, of course, is a little larger than any single episode of personal upheaval.

Nowhere else in the world has the mandate to protect and restore a species been so well buttressed by federal laws, international treaties, and high-court directives. And nowhere has that mandate been as successfully evaded through a skillfully directed symphony of public-relations scams, filthy politics, and crooked science. So many canards have been generated and dutifully reprinted that the casual tracker of this story might have a hard time discerning, as Ed Chaney might say it, where the horseshit ends and the horse begins. To dissect them all would require a treatise the thickness of a dictionary.

To give but one choice example, consider one of the oft-repeated "talking points" offered up by the BPA and its defenders: removing the four lower Snake dams would save only four of the thirteen threatened or endangered species of salmon in the Columbia. Some simple map work exposes this statement as a fraud. "Only" four species of salmon would be saved by tearing out these dams because the other nine listed species live in tributaries of the Columbia Basin other than the Snake River. Employing the same logic, one could say an international marine reserve off the coast of Greenland will do little to save Alaska's polar bears. But the claim is misleading on other levels. The Snake and its tributaries once provided half of all the chinook salmon in the Columbia Basin. The 5,500 miles of habitat now blocked by the dams represent nearly half of all the freshwater habitat available in the largest watershed on the Pacific Coast. And that habitat is in better shape and better protected than anyplace else in the basin.

"This is what makes this a hard story for an unsuspecting scientist or journalist," DeHart told me over a burger and a beer. "It forces you to figure out where your obligation to objectivity ends and where politics begin, how the latter affects your ability to do your job. All of this is confusing to your average person on the street. I can live with the kind of spin that industry puts on every kind of information. That happens all the time. But the intimidation, the censorship of science. People need to know this is the kind of situation you wind up with when powerful interests are allowed to squelch and censor scientific inquiry and debate."

Having worked on the issue since the days when delusion on the salmon issue was king, funded by an agency that, to put it mildly, has not always appreciated her work, DeHart possesses the thorough command of technical issues found in a seasoned scientist. The long shadow hydropower interests have cast over the Fish Passage Center has forged in her an even sharper sense of the politics that have affected her job. She is as open and honest about her political opinions as her scientific ones. "Bonneville," DeHart told me in her office about thirty blocks from BPA headquarters in Portland, "is not paying for science. They're paying for results that support their position. Scientists are like everyone else. They work for pay because they have kids and mortgages and car payments. And if some of your data is leading you to a conclusion that's controversial, it comes down to whether or not you're willing to kick a hornet's nest. I talk to scientists every week who tell me: 'Michele, you know I can't say that,' or 'I've been told I can't talk about that.' People are afraid to write anything down. We used to work with all agencies. We could openly disagree with their conclusions. Not anymore."

As always, the trouble comes down to money. If you want to work on salmon research in the Pacific Northwest, chances are you'll be working under the influence of the BPA's cash flow. Graduate programs in fisheries research at public universities in the Pacific Northwest are underwritten by Columbia Basin hydropower. Fish and wildlife departments in Oregon, Washington, and Idaho receive millions of BPA dollars. The NPCC, the entity responsible for creating a fish-and-wildlife restoration plan for the Columbia Basin, is wholly funded by the BPA. Monitoring and modification of fish passage systems at the dams themselves come out of BPA coffers, as do all tribal hatcheries and habitat restoration programs. The National Oceanic and Atmospheric Administration (NOAA) fisheries, the Corps, the Fish and Wildlife Service, the Pacific Fishery Management Council, even Fisheries and Oceans Canada have all cashed checks from the BPA. And the BPA maintains its own stable of private science consultants whose research invariably confirms their own ideas about the state of salmon. Single-payer salmon recovery, as DeHart sees it, has allowed the benefactor of all this science to become its own favored beneficiary. This reversal of intent in law, policy, and science has resulted in the most expensive endangered species salvage and recovery operation in American history. But instead of recovering the species, their numbers have dwindled to half of what they were when the rescue operation began. In light of the lavish budget involved, failure on such a massive scale might seem like a difficult thing to plan on purpose. But salmon recovery on the Columbia and Snake rivers has mostly been a well-coordinated effort to save dams, cheap electricity, and irrigation rather than salmon. Controlling discourse among scientists has been a cornerstone of this campaign.

Kintama Research, based in Nanaimo, British Columbia, founded by Dr. David Welch, typifies the kind of private scientific consultant on which the BPA relies. Kintama has garnered $7 million from the BPA since 2005, ostensibly to test its own brand of acoustic transponder, a tracking device implanted in juvenile salmon. In the fall of 2008, Welch published a paper reporting on the results of his work on the Public Library of Science Web site. He concluded that smolts leaving the undammed Fraser River in British Columbia survived no better than those swimming through eight dams on the Snake and Columbia rivers.

Welch's new tracking devices managed to detect only two chinook salmon in the Fraser River in 2004. Scrutiny of Welch's paper revealed procedural errors unlikely to be duplicated in even the most careless undergraduate research. Welch lost several of his paper's co-authors as they quickly hedged away from his claims. After the controversy subsided, it came to light that Welch has been extremely slow to furnish data key to verifying his work. He was a year late with an annual report required of all BPA contractors. "If I was a year late with an annual report," Michele DeHart said, "all hell would break loose."

In colloquial Japanese, as the Web site for Welch's company explains, Kintama translates to "golden balls." In making his pitch for a few million bucks to find two chinook, Welch properly figured out the language by which ratepayer and taxpayer money could be used to give his golden balls a velvet-gloved squeeze. He could prostrate himself before the BPA's corollary to the theorem that salmon don't need rivers: it's the ocean that's killing all the salmon. The premise for this contention is a phenomenon known as the Pacific Decadal Oscillation. This is a fancy term for a recently discovered twenty- to thirty-year pattern of upwelling in the Pacific that translates into periods of easier and harder years for Pacific salmon to find food at sea. Salmon populations increase in years when upwelling provides them with better pickings for food. Taking these accepted facts and running with them, some of the BPA's mercenary scientists have jumped to the conclusion that so much depends on shifting ocean conditions that what happens in rivers doesn't matter much. That idea ignores a basic fact of the salmon life cycle. "The ocean is important to salmon," acknowledged DeHart, "but without a healthy river system to deliver them there, and allow them to return to spawn, it doesn't matter if the ocean is looking good or bad." Nonetheless, when the gathered evidence showed spill to be working, some of these scientists were quick to give full credit to the ocean for the miraculous rebound. But if ocean conditions had indeed accounted for the vast majority of improved salmon returns from 2005 to 2009, where was the commensurate bump in rivers neighboring the Columbia? Didn't these fish feed from the same ocean?

Answers for the ocean argument and other curious conclusions might start with Dr. James Anderson, an assistant professor at the University of Washington's School of Aquatic and Fishery Sciences. Anderson is a modeler. He builds computer programs that represent salmon survival scenarios under different river and ocean conditions expressed in terms of statistical probabilities, a tool that can be used ostensibly to better understand what's happening to salmon as they pass through dams in the hydrosystem. Anderson's models reliably align with the BPA's contention that dedicating more water to salmon and less to power won't necessarily help the fish. It is a position that finds little support among other fisheries scientists and one made less credible by the fact that Anderson's exclusive source of funding throughout his thirty-year career has been the BPA.

Thanks to Anderson's presence, the University of Washington's fisheries program receives far more money from the BPA than any other institution of higher education, having hauled in $15 million over the past decade. Anderson has dipped into the power agency's pockets on behalf of causes other than public education. He's also cofounder of Columbia Basin Research, dedicated to salmon science on the Snake and Columbia, with a staff of twelve and offices in downtown Seattle, a venture that relies solely on BPA funds. Above and beyond these two concerns, on at least one occasion, Anderson was working on a retainer directly from the BPA. When I interviewed him in his office at the University of Washington on a blustery winter afternoon in 2009, I asked Anderson how he differentiated from among all the hats he wears—professor, consultant, scientist. "I don't differentiate," Anderson told me. "To me, they are all the same hat." Some of his colleagues would beg to differ.

In 1998, a multiagency salmon science project known as the Plan for Analyzing and Testing Hypotheses (PATH) got its start, sponsored by the BPA. A federal court had criticized the BPA for ignoring the science produced by states, tribes, and other federal agencies; PATH was the way to respond to these critiques. The idea was to bring salmon scientists, including Dr. Anderson, all around one table to identify critical uncertainties in the modeling process and, if possible, to create consensus around which management strategies held the greatest possible benefit for salmon. Thirty federal, state, tribal, and independent scientists operated under agreed-upon scientific protocols. Four scientists, chosen from outside the confines of the Columbia Basin in order to avoid bias, were picked to review the collaborative work of the thirty. PATH found the action most likely to recover endangered Snake River stocks of salmon to be breaching the dams.

Anderson was a vocal critic of the conclusion. As a scientist representing the objective view of a highly reputed research university, his opinion carried a lot of weight. It might have carried less if PATH members were cognizant of the fact that during the same years their collaborative process was ongoing, Anderson was being paid directly by the BPA, earning himself a cool hundred thousand dollars in the process. Was Anderson working as a private consultant, or an impartial university scientist, or in some as-yet unrevealed role? To at least one other scientist involved in the process, it was clear where Anderson stood.

"At the end of the PATH process," recalled Dr. Randall Peterman, a fisheries scientist at Simon Fraser University in British Columbia, who was one of four scientists named to the independent review of the study, "it was all the BPA consultants and Anderson who were saying the process was flawed. But it wasn't. The conclusions of PATH were made using the best science available, and the methods were agreed to by everyone involved. I would still stand by the results they achieved." What more the PATH group might have achieved is anybody's guess. Citing concerns that Anderson raised, the BPA pulled funding from the study as it neared completion in 1999.

For his part, Anderson denies that sources of funding have influenced his findings. "You have to look at who's doing the best science. Are their methods transparent? Are their studies peer-reviewed? You can't really say anything about a study based on where the money comes from. Someone with a political view can easily bend the data to fit their conclusions, and that's more of a thing to watch out for," Anderson told me. Because of the degree of specialization required to decipher the complex models he builds, he believes peer review is the only way to vet the viability of any claims. I asked him if complexity and transparency are mutually exclusive. "The problem with someone like you, a journalist, passing judgment on something like this," Anderson replied, tossing a stack of mathematical proofs and formulas listed in the appendices of a recent academic paper toward me, "is that you'll never be able to tell what's hidden in the numbers. Someone could make the model say what they want it to, and only a very few people would ever be able to detect any kind of bias."

Some forms of bias, however, are easier to detect than

others. Anderson has also been paid to weigh in on hydrological matters on behalf of Boeing, an agribusiness venture called Inland Land, and to Congress on behalf of power interests. Given the choice between "kicking the hornet's nest," as DeHart said it, or collecting honey, Anderson figured on sticking to the latter task.

Anyone paying attention to the nature of scientific discourse over the past decade should hardly be shocked by the kind of enterprise represented by scientists like Welch and Anderson. Like civilian casualties in modern warfare, industry-spun science is morally indefensible, but for the time being so common it does not seem subject to reform. "The BPA has been very good to me," Anderson told me. "They've allowed me to study exactly what I've wanted my entire career. Who knows? We may yet tell them something they don't want to hear."

But you could bet Anderson a tenured faculty position at the institution of his choice that his remarks won't ever offend the ears of his funding sources. Hydropower interests, as Anderson has it rightly figured, react violently to news they don't want to hear. Science consultants work for their clients. The results of their work are therefore predictable with a near-scientific precision. In response to this plot development, Anderson has pinned his career to the precarious rationale that if your numbers are good, your ethics are at least defensible. In spite of overwhelming evidence that this can't account for the entire truth, the idea is a tenet of a nearly religious faith in the private army of science consultants the BPA has suckled over the years. And like any resolute faith, its built-in blind spots make it especially susceptible to exploitation, which is what happened over the last decade with the BPA. Some familiar actors more interested in conducting political experiments than scientific ones set out to answer a question: if a few million bucks invested annually could buy the undying loyalty of dozens of scientific consultants, what could $130 million over a decade do to influence a federal research institution?

The BPA isn't supposed to be a corporation, but in 2001,

it caught a Bush-era regulatory break worthy of the best-connected Fortune 500 company. One of its own was appointed to run the very agency that oversees their dam operations to ensure they comply with federal environmental laws. D. Robert Lohn, a Montana attorney, was general counsel for the NPCC under its previous name, the Northwest Power Planning Council, from 1987 to 1994. From 1994 until 1999, he managed Fish and Wildlife for the BPA, then jumped back to the NPCC as director of its Fish and Wildlife Program. He stayed there until 2001, when Secretary of Commerce Don Evans appointed Lohn as Northwest regional administrator of the NOAA National Marine Fisheries Service.[1] True to Bush administration form, Lohn displayed an unusual skepticism toward science, even when it came at the behest of his own agency.

In 2004, one of Lohn's own NOAA scientists hired an outside team to review whether or not hatchery coho salmon should be counted as part of the endangered wild population. The scientist's opinion: hatchery fish were not wild fish. Lohn told PBS's *NewsHour*, he felt the scientists were wrong: "Their basic advice was 'the law is wrong. You should ignore it.' They're important comments, we took them as concerns, but they weren't necessarily scientific findings."

If scientists don't really produce science, then it follows that Congress doesn't really make laws. Lohn was equally skeptical of the ESA, arguing with a by now familiar twist of logic that "we end up focusing on ESA species to the exclusion of others, and I don't think that's the best way to manage." What the ESA actually says is that its purpose is to provide an impetus for the arduous work of restoring whole ecosystems. It was not written with the idea of preserving isolated species like nostalgic museum pieces.

According to BPA records, during Lohn's tenure at NOAA, the research agency took in $83 million from Lohn's erstwhile employer, the BPA. NOAA also received significant funding from the Corps: $51.5 million from 2000 to 2009. To spend this budget increase in ways he deemed wise, Lohn hired two former BPA staffers to come with him to NOAA, Rob Walton and Bruce Suzumoto. Walton heads a specially designated Salmon Recovery Division in Portland. His qualifications for this job are sketchy at best. He's not a scientist. He worked for a mining concern in Alaska and then as a staffer for a legislative committee in the State of Washington. Suzumoto has a background in fish farming and worked for the timber giant Weyerhaeuser. Both Walton and Suzumoto were employed at the Public Power Council, a utility lobby, before joining NOAA.[2]

A good portion of the BPA's largesse lavished on Lohn's NOAA went to producing a document required under the ESA known as a biological opinion. The bi-op, as it has come to be abbreviated in salmon-centric circles, is supposed to be the fruit of a laborious consultation among agencies carrying out various kinds of management plans, with another federal agency acting as arbiter—usually the Fish and Wildlife Service, but in matters pertaining to oceans, NOAA—to determine if a proposed action will further imperil an endangered critter residing in the territory in question. The ESA's blueprint for this process makes it clear that the agency with a plan to move into a sensitive habitat is supposed to *propose* measures to minimize risk; the arbitrating agency is to *render judgment* about whether this proposal will not further jeopardize conditions. This is not what has happened on the Columbia. The history of the relationship between the BPA and NOAA on this legal front offers a convincing counterpoint to critics who say environmental laws put the interests of animals above human concerns.

Since 1992, when NOAA Fisheries Service produced its first Columbia Basin salmon bi-op, courts have found all but two bi-ops they've produced to be illegal, falling short of the protection the law requires. During Lohn's time, NOAA Fisheries Service bi-ops degenerated from falling short of the law to a frivolous attempt to disregard it altogether. In 2004, the Department of Justice, on behalf of the BPA, NOAA, and the rest of the federal family, argued that dams had been around long enough that they were part of the natural landscape. Therefore, to demand changes in their operation was an act akin to turning back the tide. This was the point at which Redden effectively rolled his eyes, told the feds to get serious, and decreed that until they did, water would be spilled to benefit fish under the supervision of his court. To encourage a little more gravity on the part of federal defendants, the judge had both sides in the issue come up with a collaborative framework for producing a new bi-op, which was agreed upon in late 2005. State, tribal, and federal agencies would work together on the science behind making a viable plan. Stakeholders seemed cautiously optimistic that a scientifically sound, politically palatable opinion could be hammered out. Bob Lohn had other ideas.

With the help of fellow attorneys at the Department of Justice, in September 2006, Lohn drafted a memo that dictated to his agency's scientists the standards they should adhere to in producing an improved bi-op. Such "improvements," unfortunately, would benefit parties other than endangered salmon. Instead of planning for actions that would reasonably assure endangered species could recover, Lohn reinterpreted the law to mean that such species only had to be "trending toward recovery." This meant that in tributaries to the upper Columbia and Snake, where some stocks teeter on the edge of extinction, the addition of a single fish per year would satisfy the legal mandate to protect them. As one tribal attorney pointed out in court, under Lohn's interpretation of the ESA, salmon in certain drainages could actually become extinct (it takes two fish to perpetuate a species, and some salmon-deprived tributaries are down to one or none in some years) and still be considered "trending toward recovery." Another tribal attorney pointed out it would take several centuries for certain stocks to meet minimum recovery goals under the conditions laid out in Lohn's decree.

Even with the bar thus lowered, it was still questionable whether status quo management of the dams could achieve the minimal improvements promised in the concept of "trending toward recovery." NOAA's own empirical data on the matter suggested it would not. NOAA's preferred computer model, however, confirmed those goals would be met. And here is where salmon science runs headlong into the legacy of its loyal BPA consultants: that model is run by a former student of Anderson named Rich Zabel, now a NOAA scientist. The models of mentor and student are similar, though not identical. Zabel's model does account for some of the benefits of spill. But like Anderson's, this model has regularly been critiqued as overly complex and data-hungry, with a tendency to crunch numbers for which no correlative information exists in the real world. Zabel's model, which goes by the acronym COMPASS (Comprehensive Passage), and Anderson's model, dubbed CRiSP (Columbia River Salmon Passage), were both so routinely derided by fellow modelers as being alike that they began to refer to the NOAA model as Crisp ass.

Run prospectively, that is, as a forecasting tool to predict future results of management actions, COMPASS has struck out on several notable occasions. It didn't predict the huge losses to out-migrating salmon in the drought year of 2001, when water dedicated for fish passage was used instead to generate electricity. It missed subsequent gains in fish surviving in-river in 2007 and 2009, when the court ordered water over the dams for salmon. A favored platitude of modelers, who are now used to predicting everything from economic conditions to climate change, is that all models are wrong but some are useful. But to what end? Subject to easy manipulation, models may have as much political as scientific utility.

For his part, Zabel resents the idea that his model and Anderson's are one and the same. Despite the pressure that might be associated with being a lead scientist on a key portion of a bi-op in the highest-profile ESA case in the nation, he claims the structure of his agency allows him to be sheltered from any political storm. "I've never felt any political pressure here, and I'll tell you what, if I did, I'd quit, kicking and screaming," Zabel told me in early 2009. "My job is simply to produce the best science possible. I'm lucky enough to work for an agency that's big enough to have a science wing separate from a policy wing."

Zabel's description of laboring in a scientific haven protected from the evils of politics is one frequently heard from some NOAA researchers. This theory of objectivity, however, winds up looking quite a bit different in practice.

· In August 2005, as the federal family set out to correct the legal collision course it had set sail for with the dams-as-part-of-the-natural-environment ploy, a team of scientists from NOAA, the Fish and Wildlife Service, the states of Oregon and Washington, and the Columbia River tribes met to discuss ways Zabel's model could be made to better reflect the reality of what was happening to salmon in the river. Persistent concerns over Zabel's model had been brought to the table. A letter to NOAA signed by six biologists representing the Fish and Wildlife Service; the states of Idaho, Washington, and Oregon; and several Indian tribes laid bare a sense that NOAA was simply going through the motions of taking input from other agencies. Their concern was that NOAA's vision of collaboration did not include responding to criticisms leveled by the very team they'd asked for help. Four more times, between December 2005 and July 2006, the same scientists jointly signed letters of concern addressed to Zabel. No written response ever came back.

According to the Fish and Wildlife Service's fish-passage modeler, Steve Haeseker, it didn't matter if such concerns were voiced during face-to-face meetings or in writing. "There was never any effort to address the concerns once they were raised," Haeseker told me. "It was pretty clear NOAA had already decided what the model was going to look like. And if they were really interested in collaboration, there's a better way to do it than just reviewing different versions back and forth. The best way to improve on a model like this anyway is to share the code [the computer program for the model]. We asked NOAA for the code to COMPASS early on in this process, and the answer was no."

By summer of 2006, it was clear Zabel's model would figure prominently in the court-ordered revision of the bi-op. Those same six scientists reiterated their long-standing concerns. The

Fish and Wildlife Service and other agencies took the extraordinary step of requesting their names be removed as coauthors of the revised manual for COMPASS. Zabel obliged, then redirected any criticism to be sent directly to a review board, the Independent Scientific Advisory Board (ISAB), created to quell some of the controversies surrounding regional salmon science. They issued their verdict six weeks later. Zabel and his staff were admonished for going at the model unilaterally and encouraged to incorporate the concerns of fellow scientists in future rounds of modeling revision. The finding prompted Suzumoto, the former utility lobby staffer and Lohn's right-hand man, to do two things. The first was to call up Dan Diggs, assistant regional director over at the the U.S. Fish and Wildlife Service. The second was to get Zabel on the task of writing up a report covering the whole affair to date.

After Suzumoto called Diggs, the latter quickly wrote his employees who'd been working on Zabel's model. He informed them that Fish and Wildlife Service scientists were being pulled from modeling work. His rationale for muzzling his own scientists provides a rare glimpse into the forces by which politics trump science in the effort to save salmon:

> While states and tribes and even independent science panels may support our perspective at times, it is critically important that we be able to present and persuade the federal decision-makers. . . . Final decisions in these hydro issues are those of other federal agencies with huge biological, legal, policy, and political implications. If we are viewed as "outside the process" or renegade (rightfully or wrongfully) in won't matter how good our science is, it will be ignored, which unfortunately is what is happening to date apparently [*sic*] in the Compass modeling exercise.

"The best available science," the standard called for in the prevailing federal law, was easily eclipsed by politics, as Diggs called it.

In the aftermath of the flap with Fish and Wildlife Service, it became clear that NOAA isn't really dedicated to keeping science separated from politics like oil from water. In early 2007, *New York Times* reporter Felicity Barringer got hold of the critique of NOAA's model that Fish and Wildlife Service scientists had sent to the ISAB. When she contacted NOAA Fisheries Service for a response, Lohn e-mailed Zabel asking for help in answering the reporter's inquiry. After a quick confab with Zabel, Lohn pointed Barringer to Zabel's report, adding that "there was no attempt to preclude the [Fish and Wildlife Service] technical staff's views."

This was a bald-faced lie. Refusing to share the programming code, stonewalling persistent critiques, and responding to recommendations from an independent panel by conniving to have dissenting scientists pulled from a project hardly exemplifies the spirit of free inquiry. Moreover, Zabel's efforts on behalf of his policy boss in fending off the opinions of colleagues and the inquiries of reporters place him squarely inside the policy side of NOAA. Yet there was worse to come: about this time, a new variable was being introduced into Zabel's model. An employee of the BPA began making the rounds of NOAA scientists working on the bi-op. If scientists were helping out on the policy side, it was only fitting that a strict policy man would be there to muscle in on the science.

Jeff Stier is the BPA's senior policy advisor for Fish and Wildlife. Like Suzumoto and Walton at NOAA, Stier isn't a scientist. He started out in politics as chief of staff for Congressman Peter DeFazio (Oregon). From there he jumped to the BPA, toiling in bureaucratic anonymity out of their D.C. office until 2005, when he accepted the BPA Fish and Wildlife job and a move back to Oregon. Now in most organizations, a man with no experience or qualifications in matters pertaining to fish and wildlife affairs would not wind up second-in-command in that very department. But most organizations don't operate like the BPA. Lohn had lowered the standard in a key legal provision of the ESA; Stier would work on the science end of things, helping to make sure the numbers affirmed the wisdom of Lohn's directive. Records of Stier's constant contact with NOAA scientists from 2005 to 2009 reflect the considerable pressure he put on them to lower the standard for various types of data analysis and measurement. Wherever inconsistencies or uncertainties in the science could be detected, Stier would run in to claim these considerable gray areas, asserting early and often that when no clear line could be drawn from hard data to specific recommendations, the standard should automatically revert to what he deemed "a policy call." Instead of making carefully informed choices based on the weight of available evidence, decisions were made based on the BPA's political preferences. Thus the federal family could "prove" that a biological opinion written under the guiding light of "trending toward recovery" wouldn't cause further harm to endangered salmon.[3]

In a telephone conversation with me, Zabel declined comment on Stier's role in getting his model to affirm the BPA's bi-op. But later, I spoke with Zabel and John Ferguson, NOAA scientists with roles in composing the current salmon biological opinion. Both told me unequivocally that spill boosts salmon survival rates. Why, then, does the latest bi-op *call for less spill than what Judge Redden's court has mandated since 2005?* "I am an environmentalist," Zabel told me. "If it were only up to me, we'd be doing everything we could to bring these fish back, including tearing out dams. But I'm also a scientist. I can only go where the numbers take me."

"You just can't wear both the policy and science hats into the same meeting," Ferguson chimed in. "What you want and what the data says are bound to get mixed up. We're all biased. There's lots of psychological studies that show people looking at just the raw facts on any given subject generally find what they're looking for. I can't let that happen as a scientist." But the pursuit of pure objectivity has its own hazards. Input equals output: suppose nothing, and you're inclined to find nothing. Into that void rushes the industry representative, the political agent, the prostituted scientist. Unlike the objective scientist, these people have always known what they want the numbers to express: that there is no reasonable or desirable limit to growth, expansion, or profit. I asked Ferguson what his feelings were about having Jeff Stier at the table with NOAA scientists as they hammered out the bi-op. "I have no problem with him being there," Ferguson told me. "He understands a lot of the issues, and the BPA has a lot invested in this process."

Suffice it to say, Stier found what he was looking for. By contrast, objective scientists traveling only to where the numbers lead were turned against the values they wish to uphold.

In one way, it doesn't matter how effectively Stier managed to sabotage the deliberations of NOAA scientists. With the help of a science consultant working for the BPA, Stier wrote significant portions of the 2008 bi-op, a fact he uncharacteristically let slip as the BPA was getting ready to pitch their Bush-era salmon plan to the incoming Obama administration in April 2009:

> Let's talk about who does what after we talk to Steve [Wright]. . . . Sadly the part of the briefing where I could add the most value is NOAA's part-including the status issues, but also anything having to do with the analysis, the underlying science, etc. Maybe we should try to get me attached to their effort somehow. For instance, the extinction risk analyses are ours (me and Hinrichsen).[4] The analytic approach is ours.

How did Jeff Stier get a seat at the table with NOAA's scientists? His employer paid for it. Stier's expressed pride of ownership in NOAA's science is backed by the numbers. Since 2001, the Northwest Fisheries Science Center, the Seattle office of NOAA charged with writing biological opinions for salmon in the Columbia Basin, has received more than three-quarters of its budget—$90.2 million—from the Bonneville Power Administration ($38.7 million) and the U.S. Army Corps of Engineers ($51.5 million). This arrangement places NOAA and the Corps in an ethically compromising position. The two agencies are named as federal defendants in the bi-op court case, with the

BPA offering its full arsenal of legal and financial support to the defense. To draw an analogy that addresses only the most immediate conflict of interest, say you are in a contract dispute with your boss. A judge orders your case into arbitration. The mediator recommends you move closer to your boss's position. You find out the mediator is being paid by your boss. You call your lawyer.

~ ~ ~

The scientific community did not hold Jeff Stier's debut as a fisheries scientist in high regard. More than a hundred actual scientists signed a letter expressing concern over the 2008 biological opinion, and a 2009 set of amendments to that opinion, parts of which were also written by Stier. They observed that the '08 opinion provided less protection for salmon than previous ones rejected by the courts in years past. Several of the biological "triggers" that would trip additional protection if salmon counts declined were set lower than the alarming numbers that got salmon stocks listed as endangered in the first place. The American Fisheries Society, a consortium of professional fish biologists, also issued a fifty-page report, finding the federal salmon opinion to be "inadequate for ensuring the protection of threatened and endangered salmon and steelhead in the Columbia River Basin" and concluded it did not consistently use the best available science.

But help, it was incorrectly assumed, was on the way. In his inspiring inaugural address, Obama had promised to "restore science to its rightful place." Again, in April 2009, speaking before the National Academy of Sciences, Obama announced a new executive order mandating an open-door science policy at federal agencies. He proclaimed, "The days of science taking a back seat to ideology are over."

His choice for director of NOAA was a world-renowned Oregon State University marine ecologist, Dr. Jane Lubchenco. In a speech in Denver in early 2009, she echoed her boss's calls for honest science. "Science will be respected at NOAA; science will not be muzzled," she declared. "We won't change or distort scientific findings to meet any preconceived ideology."

In the spring of 2009, conventional wisdom had it that Obama was poised to pull the plug on the Bush administration's bi-op. In May, Judge Redden wrote a letter to both sides in the bi-op case, reiterating his skepticism of the federal defendants' claims. Federal defendants, in turn, asked the judge for a six-month extension for the final draft of the biological opinion over which Redden would deliberate. Redden granted the request. But the last thing the feds deserved was more time.

Freedom of Information Act requests made of the BPA, the Corps, NOAA Fisheries Service, and the White House Council on Environmental Quality (CEQ) produced a record that shows the federal family was not using the extension of the deadline granted to them by the court to work with the plaintiffs to address the concerns laid out by Redden. They were getting ready instead to convince the Obama White House that the Bush-era bi-op and salmon recovery plan represented "the best available science."

The federal family had already as much as admitted it was not going to change course. In April 2009, federal defendants, trying to fend off any changes to the bi-op the Obama administration might be considering, emphatically vowed to Redden that they "cannot and will not amend the final agency action that is currently the subject of this litigation" and had "no interest in reinitiating consultation on the issue." To that end, in April 2009—a few weeks before Lubchenco was all ears in Portland—CEQ attorneys, BPA and NOAA staffers, and Corps and Bureau of Rec personnel jointly presented a slide show to Obama administration officials that was a rerun of many of the federal family's standard canards. The federal family claimed the process of creating the 2008 bi-op was both fully collaborative and transparent. They flashed graphs that depicted the high costs of spill and dam removal. They used their own inflated and outdated numbers for these costs, ignoring fresh calculations for replacing power lost to dam removal that showed a free-flowing Snake was economically feasible. They tossed in a novel graphic that showed how modifying flows at dams to make them more fish-friendly contributes to greenhouse gas. They claimed it would take three nuclear power plants to replace power lost from the removal of four Snake River dams. No mention was made of Redden's consistent and long-standing concerns over flaws in the plan the federal family was pitching to the Obama White House. In May of 2009, Jane Lubchenco arrived in Portland, Oregon, for what was billed as a "listening session" on salmon matters in the Columbia Basin. She introduced herself to a select group in a closed-door meeting by saying, "We are here to learn from the past so we can do justice to the future." Lubchenco may not have known it then, but the course for the future had already been set to steer clear of justice.

When President Obama tapped former Washington governor Gary Locke to be his commerce secretary, the State of Washington was unwittingly granted extraordinary power in the legal and political wrangling over salmon. Washington gets almost all the irrigation benefits from the Columbia River dams, and buys two-thirds of its power output. As governor, Locke learned well what every politician from his home state does to maximize the odds of staying in office: loyalty to status quo dam politics on its southern border. Locke also retained some cozy relationships to home that have galvanized his state's opposition to meaningful hydropower reform. His top aide is the daughter of Washington governor Christine Gregoire. Daughter Courtney jumped to Commerce from the staff of Washington senator Maria Cantwell. Cantwell chairs the Senate Coast Guard and Fisheries Subcommittee, which directs funds for much of NOAA's work. Locke, Cantwell, Governor Gregoire, and senior Washington senator Patty Murray—Democrats all—each took part in a tacit effort to stump for a biological opinion that falls well short of the scientific standard President Obama called for.

Freedom of Information Act records show that Lubchenco's restricted visit was orchestrated by Locke, Cantwell, and

Murray. The senators' respective legislative staffers assigned to the issue, Murray's Jaime Shimek and Cantwell's Joel Merkel, worked feverishly before Lubchenco's Portland visit to schedule a meeting with their bosses, Nancy Sutley, CEQ chair, and Terrence "Rock" Salt, principal deputy assistant secretary of the Corps' Civil Works Program—the latter an influential figure in one of the agencies named as defendants in the bi-op case.

Simultaneously, Merkel and Shimek laid out the itinerary for Lubchenco's Portland trip. They recommended she meet "only with sovereigns" (that is, no environmental groups) and that she limit discussions to what could be accomplished only via the anemic jeopardy standard[5], reminding her that "recovery is a broader issue." They advised she be "discreet" and "avoid open public meetings," as well as groups that had "gone political" on the issue. Meanwhile, they recommended a list of highly political key personnel to visit—regional officials from the Corps, scientists working as consultants for hydropower interests, lawyers working for groups advocating the Bush bi-op, and members of the NPCC known to take a dim view of any ambitions on behalf of salmon that take water away from industrial or agricultural concerns. Shimek, Merkel, and Courtney Gregoire checked in on Lubchenco early and often, Shimek at one point fuming that the amount of time Lubchenco spent with plaintiffs was becoming "disturbing." She needn't have worried. Staffers from Locke's Commerce offices e-mailed the senators' offices, writing they had heard the senators' concerns "loud and clear." They developed an itinerary "with which [the senators] would be comfortable." The trip took place as the Washington delegation had suggested.

On June 4, 2009, Lubchenco met again with the Washington delegation, after which a series of op-eds and public appearances, in which she offered her support of the Bush bi-op, began to materialize. Colleagues and friends in Oregon wondered how one of the nation's premier scientists could have been so quickly muzzled in her baptism into national politics.

The man who could probably best answer that question was Jane Lubchenco's boss, Gary Locke. While Lubchenco listened, Locke talked. On July 23, 2009, weeks ahead of the deadline for NOAA scientists to file their latest opinion on the bi-op, Locke phoned Oregon governor Ted Kulongoski, revealing to him that the Bush plan for saving salmon in the Columbia would become the Obama administration's. Ironically, that call was supposed to be part of the Obama administration's effort to get to know better the shortcomings of the Bush salmon plan. Kulongoski and his staff were told Locke would listen to Oregon's suggestions for specific ways that the biological opinion could be improved.

With powerful people in just the right places, it became possible for the State of Washington to not only maintain the status quo but benefit from its entrenchment. By the end of 2009, the state, which had remained quietly supportive of the plaintiffs in the bi-op case, signed on with the federal defendants. As a reward for adhering to this lower standard, the Washington Department of Fish and Wildlife cashed a $40.5 million check from the BPA for habitat enhancement in the Columbia River Estuary, a location aligning well with the industry belief that problems for salmon have nothing to do with dams.

Cutting these kinds of deals is not without risk. In June 2009, the Department of Justice made some preliminary inquiries into just what was said at a meeting among Governor Gregoire, Locke, and representatives of Washington's agribusiness industry when they all gathered to discuss the bi-op.

While Commerce did its part, the Washington senators went to work on their colleagues. When Senator Jim Risch (Idaho) made overtures toward facilitating a roundtable discussion to reach regional consensus on the salmon issue, Murray castigated members of the Pacific Northwest delegation to make sure they did not get behind Risch's effort. Thus dysfunctional salmon politics have turned conventional political alignments inside out: a prominent Western Democrat was riding herd on a Republican to stifle progress on a key environmental issue.

At least, Risch seems to know how to keep the fun in dysfunctional. "This whole issue is personality-driven," he told me in a
phone conversation in early 2010. "As soon as we have a critical
mass of people who can overcome that, maybe we can get together in Congress, hold hands and sing 'Kumbaya' and come
up with a real solution."

Few who track the work of the Senate these days are holding out for a "real solution" to any of the nation's pressing
problems, let alone one as far down the list as some dams in a
relatively unpopulated corner of the country. Nonetheless, high
crimes and misdemeanors committed against the river, salmon,
and the people who depend on them will have far-reaching
consequences.

The micromanagement of Dr. Lubchenco, part of the
larger effort to hawk a Bush salmon plan to the new administration, was well underway before the ink was dry on President
Obama's executive order to open doors and turn on lights at
federal science agencies. If the White House was fully cognizant of the scheme of the Washington senators and Secretary
Locke, Obama's commitment to open science gained an immediate credibility problem. And if the White House didn't know,
then the president and his inner circle should probably be wary
of a commerce secretary apparently more willing to go to work
for senators from his home state than to uphold the policy of
the man who appointed him.

Over in the Senate, things don't look much brighter. Senators Patty Murray and Maria Cantwell appear to be the willing
heirs to Larry Craig's role as guardian of the dams and the fat
subsidies they represent for the region. In their hands, the fate
of the largest river in the West isn't a complex natural resource
issue worthy of wide-open scientific investigation and public
debate. It's a source of easy favors by which angry constituents
or lavish campaign contributors might be painlessly appeased.
Each has used her respective influential position to censor, stifle,
or cut short the depth and breadth of the scientific process. The
works wrought by these senators typifies the black plague afflict-

ing modern politics. Until the cure emerges, political solutions to social problems will remain anemic at best. The trouble boils down to this: instead of serving the country, the new corporate face of governance sees the myriad crises facing the nation only as a set of public relations riddles to be solved. In the future, the vast punishment and ruin this arrangement has instigated will be painful to face, much less overcome. In the meantime, earnest politicians and engaged, concerned citizens are treated like the unwitting customers of a fraudulent mail-order operation seeking redress for the bill of goods they were sold.

In December 2009, BPA administrator Steve Wright saw fit to celebrate these developments. As the keynote speaker at the annual banquet for the hydropower lobbying group Northwest River Partners, Wright first offered congratulations to his constituents for their hand in the current bi-op. He referred glowingly to "the close relationship between this new administration and River Partners" and to the bi-op, which he proclaimed should be celebrated "for its fidelity to the science, its allegiance to the law, and its adherence to meaningful collaboration." He then quelled remaining concerns over any possible wrench Lubchenco might throw into their well-executed plan. With a nod to the composition of his audience, Wright added with a smirk: "I wondered how she would operate in the policy world, where I've been for thirty years. Well, I had a chance to visit with Dr. Lubchenco last October. I'm pleased to report she gets it. She knows decisions have to be made. I found her to be resolute, firmly committed to the plan, and, interestingly enough, she really wants to win." Wright credited Northwest River Partners for clicking on the lightbulb above her head. "Thank you for your support, advice, counsel," he told the audience. "You got us to where we are."

And just what, then, is our precise location on this river of progressive enlightenment to which Wright referred? The strongest provision in law mandating salmon recovery, the Northwest Power Act, is a dead duck. None of its fish and wildlife restoration goals have ever been attained, nor has a viable plan ever been crafted to attain them. All the legal wrangling now is fought over the Endangered Species Act, which has come to emphasize merely preventing salmon from going extinct. As if that was not enough, Bob Lohn connived to have the standard by which this minimal effort would be measured lowered further still.

The Salmon Recovery Division Lohn helped create is still being run by former shills for the BPA and the utility industry. And just in case the old NOAA salmon scientists deviated from the course set for them, a BPA policy analyst wrote significant portions of the bi-op Wright so sincerely deemed true to science, law, and meaningful collaboration.

The rest of the country continues to subsidize cut-rate hydropower, at a cost of $300 million a year. That figure excludes legal costs, the greatest of which undoubtedly are yet to come. If salmon cease to run, American taxpayers will be on the hook for as much as $30 billion in damages, the amount their government will be mandated to pay for abrogating treaties with river Indians. If the prevailing political winds continue to blow from the usual direction, it might be wise to start putting this money aside now.

Northwest ratepayers are on the hook for both the federal family's multibillion-dollar debt load, as well as the cost of a salmon restoration scheme that benefits the future of dams more than salmon, now running an annual tab of more than half a billion dollars annually, all for a fix that doesn't work to repair salmon runs.

A much cheaper option could be made a reality. One study purports dam removal will save federal taxpayers $5 billion and generate more than $20 billion in revenue for the region. But getting rid of them will require an as-yet-unseen commitment to change the standard political order. And the federal family seems determined that the rest of the country won't get off so cheaply.

As a final touch in the transition from Bush to Obama, the BPA's fellow defendants in the bi-op court case, the Corps, killed any wild ideas about a serious new look at Snake River Dam breaching. As the federal family was finalizing some minor changes to the 2008 bi-op, made to give the appearance that the reprieve granted them in the summer of 2009 was time well spent, Brigadier General William E. Rapp turned down an offer from a BPA attorney, Lorri Bodi, to write the section that addressed the future of the Snake dams. Rest assured, Rapp was on the dam defender's side. He suggested his scientists include one paragraph on the matter, commenting he didn't want the science of dam breaching to get "too wide open."

Sometimes even a strictly objective scientist can see the color of money in the weave of scientific discourse. In the fall of 2009, John Ferguson at NOAA gave me his cool assessment. He was willing to admit the funding milieu was less than ideal. "It would be preferable," Ferguson said, "if we were base-funded through Congress or something; it would be great, really, to not have to try to keep a lot of scientists working year after year based on soft-money contributions. But the reality for now is different. The truth of it is that we all—every scientist working on salmon in the basin—work for the BPA to a certain degree."

By that definition, everyone at the Fish Passage Center, including DeHart, works for it, too. DeHart keeps a reminder of a quite different interpretation of this commitment above her desk. It's the letter informing her of termination as a federal employee, sent shortly after Craig connived to kill her center's funding. The brush with sudden death has not altered her conception of the center's mission. "We are seeing huge benefits from spill," DeHart said. "The BPA should actually be happy about this. Because we've done pretty much all we can in terms of modifying and improving at the dams. And the law is very clear: we cannot allow these fish to go extinct. We should be spilling as much as possible. Because if that doesn't work, the only other option is to start tearing out concrete."

The Fish Passage Center, she acknowledged, was kind of a glitch in the grand techno-industrial command into which the BPA has morphed. "I remember the day in 2005 that I was packing up this office," recalled DeHart. "And I remember thinking, This is OK. Because the one thing that follows you around your whole life, the thing you have to deal with when you look in the mirror each morning, is your integrity. We were invented to answer a question: do salmon need water? The science says salmon do enjoy water after all. For us, providing that answer has been a matter of honor."

10

A RIVER RESUSCITATED

*Salmon, Shad, and Alewives were formerly abundant here,
and taken in weirs by the Indians, who taught this method to the
whites, by whom they were used as food and as manure, until the
dam, and afterward the canal at Billerica, and the factories
at Lowell, put an end to their migrations hitherward. . . .*

*Still patiently, almost pathetically, with instinct
not to be discouraged, not to be reasoned with, revisiting
their old haunts, as if their stern fates would relent, and
still met by the Corporation with its dam. . . . Who knows
what may avail a crow-bar against that Billerica dam?*

*At length it would seem that the interests,
not of the fishes only, but of the men of Wayland,
of Sudbury, of Concord, demand the leveling of that dam.*

—HENRY DAVID THOREAU,
A Week on the Concord and Merrimack Rivers

To many observers, the lock on meaningful restoration measures owned by the federal family on the Snake River looks
impossible to break. But on the other side of the country, dam-
removal advocate George Viles can talk about how quickly activism can pry loose the grip on control. He can also talk about how
dam removal on his home river now looks like the commence-
ment, rather than the culmination, of a restoration effort. He'll
also tell you he didn't much care for the idea himself a decade
ago. Viles used to be a skeptical environmentalist. His family's
roots on the Kennebec River in southern Maine go back nine

generations. He's lived for thirty years with his wife and two daughters in Sidney, about five miles upstream from the state capitol of Augusta, right on the river. He likes it there. His acreage goes down to the river's edge. When his girls were young, they could launch a canoe into the river, and the gentle lapping of the still river against the hull would put the kids to sleep in a matter of minutes. Paddling half a mile up to the mouth of Seven Mile Brook, they would watch eels and turtles. The bass fishing was fine. He'd grown accustomed to things as they were, including the dam right in downtown Augusta that had backed up the Kennebec past his house since the mid-nineteenth century. When the decision to tear down Edwards Dam was made, he was apprehensive to say the least. "I thought it might be a good public thing but not so much a good private thing for us," Viles recalled. "We were concerned about mud and brushy banks and losing access to the river. And we thought that small towns like ours, Sidney and Vassalboro, were being ignored or overwhelmed by what people in Augusta wanted."

Worse for Viles, there was no escaping the subject. It dominated his thoughts at home, and it was a prevalent topic of discussion at the offices of the Maine Department of Environmental Protection, where he worked. "The two lead biologists for the department are still friends of mine," he said. "And they would tell me they weren't positive, but they had a pretty good sense I was going to like what was going to happen to the river."

With equal parts resignation and curiosity, Viles made the short drive downriver to Augusta on July 1, 1999. Secretary of the Interior Bruce Babbitt commemorated the event. "Today with the power of our pens," Babbitt declared, "we are dismantling several myths: that hydro dams provide clean, pollution-free energy, that hydropower is the main source of our electricity, that dams should last as long as the pyramids, that making them friendlier for fisheries is expensive and time-consuming." The removal of Edwards, Babbitt continued, was "a challenge to dam owners and operators to defend themselves—to demonstrate by hard facts, not by sentiment or myth, that the continued operation of a dam is in the public interest, economically and environmentally."

Viles didn't pay much attention to the secretary. He was more interested in watching the dam go. Bells in church towers around Augusta rang out at the appointed hour of the dam's blowout. "The first thing you thought was, how is this massive structure going to come down?" Viles recalled, standing with me at the erstwhile site of the structure along the west bank of the Kennebec in downtown Augusta, a month after the tenth-anniversary party for Edwards' removal. To his mind, back then it was all quite swift and orderly, even a little anticlimactic. Besides, the real test would be to see what the river looked like at his house. "We used to be more or less at the center of the impoundment," Viles told me. "We were told we were going to lose fifteen to twenty feet of depth right out in front of our property. I was just worried the river would be inaccessible or too muddy and smelly to want to float or fish in anymore."

Edwards Dam had blocked seventeen miles of the lower Kennebec River since 1837. It was designed to provide power to run sawmills and gristmills, and improve navigation in the river above the dam. As early as 1834, Maine citizens protested that the dam would decimate fish populations. Early New England laws required mill owners to provide fish passage at their dams. As would be the case in the West a century later, the law was generally ignored. On the Kennebec, stocks of alewives, shad, and Atlantic salmon collapsed. Fishing became a thing done further upstream or, better yet, in Canada.

In the twentieth century, Maine became the country's paper colony. Even today, a significant portion of the nation's households unwittingly wipe their backsides with the processed fiber of Maine's forests. The paper industry, in turn, used Maine's rivers as an industrial-scale bidet. The river got progressively uglier as the years rolled by. During the decades that pulp and paper mills proliferated alongside dams, sawdust smothered the river bottom and choked oxygen from the water, dioxins used in the manufacturing of paper poisoned

it, and various combinations of industrial effluents finished
the job. Long before the dam was a century old, the river was
utterly uninhabitable. Help came in the form of a spate of environmental reform laws in the 1970s and early 1980s. No sane
human being would set foot in the Kennebec even then, but
enough life was breathed back into its waters that a few wildly
hopeful anglers began plying the river for whatever fish could
survive there. Shad and Atlantic salmon were a memory, but
alewives—a variety of herring whose adaptations toward survival are predicated on the idea that there's safety in numbers—
were the first to make a comeback. Alewives are not a pretty
fish, as the name intimates. Origins of the fish's appellation may
be crude homage to the grumpy visage and ample girth of the
stereotypical tavern keeper's better half. But they once returned
in such plenitude to rivers from Florida to Canada that they
formed the base of a food pyramid that runs the gamut from
striped bass to seals and porpoises.

Alewives, some fishermen reported in the early 1980s,
had come back, milling about at the base of Edwards Dam.
Realizing their biological significance, in 1986, the Maine
Department of Marine Resources began an ambitious alewife
restoration project. Kennebec Hydro Developers Group, a
consortium of seven upriver dams, signed a fisheries restoration agreement with the State of Maine aimed at bringing back
alewife, American shad, and Atlantic salmon. The owners of
Edwards Dam, however, refused to sign on. Their license with
the FERC was going to expire; they weren't about to go on the
hook for several million dollars of fish-passage improvements if
FERC said they didn't need them.

Some of the fishermen that had first publicized the alewives
holed up below Edwards began to say publicly that the dam's
time had long since passed. Edwards' purpose in its later stages
was to power a textile mill that provided some eight hundred
jobs. In the early 1980s, these workers were permanently fur-
loughed, and suddenly the dam no longer had an economic justification for its existence. A group calling itself the Kennebec

River Anglers Coalition began floating the idea that the dam should go. The modest proposal of these pissed-off fishermen was that those responsible for ruining the river should pay for the decommissioning of Edwards.

A fire in 1989 extinguished any hope of the textile mill ever reopening. But power-generation facilities damaged in the fire were repaired. Generating capacity wasn't much—a mere 3.5 megawatts—but the dam's owners found a sweetheart deal with Central Maine Power, signing a fifteen-year contract to put power onto Maine's grid for about three times fair market price. Electricity sales, the owners found, could also be used to curry favor with the municipal government of Augusta. With a fight for relicensing in the offing, Edwards Manufacturing inked a deal to become co-licensee of the dam along with the city. For their trouble, Augusta got 3 percent of the power revenues. But the State of Maine was behind removal. The Kennebec River Anglers Coalition shortened its name to the Kennebec Coalition but added some clout from a few powerful national conservation groups, including Trout Unlimited and American Rivers. The balance of political power was for once on the side of the river. But FERC,[1] the powerful entity responsible for issuing operating licenses to the nation's privately owned dams, could've trumped regional political momentum any day.

Before 1994, FERC had rarely, if ever, met a proposal for dam relicensing it didn't like, recommended for removal. But the agency evolved to reflect the mandates of environmental law. Over the objections of Augusta politicos and Edwards owners, FERC declared in that year that it did indeed have the power to deny a license and to order a dam decommissioned at the owner's expense. The FERC then vacillated on whether to exercise that power at Edwards. At first, in 1996, it put forth a plan for fish passage, at a cost to Edwards' owners of $9 million. The Kennebec Coalition filed voluminous comments on the plan, and in 1997, with a final Environmental Impact Study in hand, FERC reversed course and decided Edwards should come down. The owners still objected to the cost. In an innovative solution to funding deconstruction, a shipyard downstream, the Bath Iron Works, along with the cooperating coalition of hydro operators upstream, agreed to finance the demolition of Edwards, implement fish passage at upstream dams, and fund fish restoration, in exchange for a grace period in meeting higher standards for water quality and accommodating fish. It was all over but the crying. The first dam in U.S. history to be removed against the owner's will was a done deal.

A decade after the fact, the river on which George Viles feels most at home has so thoroughly assuaged his fear over the Edwards Dam disappearing that he was asked to speak about his conversion to the cause at the tenth anniversary celebration. A few of the reporters that had hung on Babbitt's every word in 1999 were now awaiting Viles' pronunciations. Culling statistics from his biologist friends, whose hunches about the river proved to be correct, Viles shared what he saw in the days following demolition:

> Two months after the dam removal, the biologists' rock trap counts showed a three-fold increase in the number of different species . . . invertebrates . . . in the water and a whopping thirty-fold increase in the abundance of critters. This is the health of flowing, oxygenated water. The river smells great. The river attracts all kinds of life including paddlers and fishermen and women and those of us of all ages compelled to skip rocks. The financial, natural, and emotional value of the new river and the whole Kennebec just goes up and up. The evolving story of the return of the sea-run fish is terrific and may be the best ongoing biology lesson for young and old alike. There's no question that a free-flowing river works at cleaning itself, but with us, the Kennebec has a lot of partners.

Those biology lessons to which Viles referred are turning out to be an interdisciplinary exercise. The uncovered banks of the Kennebec fostered the return of many riparian species, including eagles, osprey, and deer. But people came back to the river along with the animals, occasionally to discover a connection to their own history they'd all but forgotten. In the months after the restoration, Viles occasionally saw savvy human scavengers combing the riparian habitat of the river and wondered more than once what they were looking for. It didn't take too long to figure it out. "The spring after the dam came out, our daughter was home from college," Viles recalled. "She'd been working at a bird refuge and brought a friend home who she'd met there. He was a herpetologist and was up at the crack of dawn every morning, down in the river. He found a spearhead that was maybe four inches long. They took it to the state museum and matched it against what they had there. It could have, they figured, come from Greenland or further across the Atlantic."

In addition to piquing curiosity on a local connection to history, further serendipity for Viles has come in seeing the river as a worthy link to his personal and familial past. "My dad was raised on the upper Kennebec for a while when the fishing was good," he reminisced. "It sounds bizarre to say it, but when I'm on the river, sometimes I think of a photo of my mother's mother when she was a kid, all dressed up in petticoats, her twin brothers in their Tyrolean outfits, all three of them in the living room of their house, all with huge rifles laid across their laps. It was still all wild country back then."

Turning back the clock on the river, according to studies published by Bates College economics professor Lynne Lewis, has sped up the present-day local economy. Property values adjacent to the former impoundment have risen. People are not only spending more time but more money fishing and floating on the Kennebec. Waterfronts in towns along the river, like Gardiner, Augusta, and Waterville, have been enhanced and restored. In Augusta at the site of the old dam, the remaining buildings of the old mill complex on the east side of the river are being demolished. In their place, city planners are envisioning some sort of mixed-use urban development complex with the rejuvenated river as its focal point. "It's funny in a way, I guess," Viles began as we watched a giant mechanical shovel wrecking part of the old mill complex across the Kennebec from where we stood. "People knew for a long time what Augusta and the Kennebec was about," he recalled, nodding toward the mills. "That's gone now, and we really don't know what the economy of the area will be based on next. But it looks like a healthier river is going to be a part of it, and I think that makes a lot of us very happy."

Happiness arrives for others as civic and cultural traditions long dormant are returning. On the green lawn between Old Fort Western and City Hall in downtown Augusta, I sat down on a muggy July afternoon at a picnic table to talk with Richard Lawrence. He's the new alewife warden for the town of Benton, some twenty miles upriver of Augusta, the first person to hold that position in 180 years. In the pre-dam days two centuries ago, the alewife warden was the man who decided when up-river commercial fishing operations could commence. "It's a traditional office that can be almost anything a town makes it," Lawrence explained. "In some towns near the coast, it's been the person who became a tycoon, the *generalissimo*. A lot of power and a lot of resentment. Sometimes the job was a paying one, but I'm not in this for the money. I saw this was something that was done before the dams went in, and it seemed like a good tradition to bring back."

Determination to maintain the democratic virtues of rural New England life seems to have been the driving force behind his efforts to put Benton's fishing rights back to work. "Small town life in Maine has changed over the past forty years," Lawrence told me. "When we first bought our land, now about 250 acres up a tributary of the Sebasticook, no one around here had much religion, but we all went to the little Catholic church in town. There was a community church and a community school; there was an annual luncheon that was the event of the year. None of that happens anymore. Here is something that used to happen in this town, and so many things like it are gone forever. When the alewives are running, the eagles and the osprey and the cormorants are out there, but people come out and watch, too." Catching alewives, Lawrence figured, might give his fellow Bentonites further reason to gather by the river. In good years, it might also swell the town's coffers a bit.

From the time Edwards was blown until May 2009, Lawrence had been planning to make Benton's first commercial harvest of alewives since Martin Van Buren was president a reality. As part of the accords signed in 1999 to bring down Edwards, fish passage had to be provided at dams upstream. Betting on success, Lawrence applied for the town of Benton to be issued a permit for the commercial harvest of alewives, based on fishing rights that had gone dormant for almost two hundred years and contingent upon the fish hitting a benchmark for returns that would allow a significant catch. He didn't have to wait long.

Fort Halifax Dam, at the mouth of the Sebasticook downstream from Benton, was the next blockade upriver from the erstwhile Edwards. As planned, Halifax had been outfitted with the means to get fish over the dam. The fish passage provision was a vast improvement over none at all. More than four hundred thousand alewives recolonized the Sebasticook in 2008. But Fort Halifax was a small, low dam. Its owners, Florida Power and Light, decided it didn't generate enough power to offset the costs of maintaining the fish passage system. Over the vehement objections of a few who owned property along the reservoir, the dam was torn out in July 2008. The response of the alewives was overwhelming: 1.8 million returned to the Sebasticook in spring of 2009. "We harvested four hundred thousand of them," Lawrence told me, "and earned about twenty thousand dollars for the town." The alewives in turn caught lobster. Maine's lobstermen baited their traps with Sebasticook alewives for the first time since the Whigs were a national political sensation.

The exponential bump in alewife numbers has lent momentum to statewide restoration efforts on behalf of the species, which is why Lawrence came to meet me in Augusta from his home outside Benton. Somewhat like sockeye salmon, alewives

rear their young in freshwater lakes and ponds until the ju-
veniles are big and strong enough to obey the call to migrate
oceanward. In coastal Maine, many of these ponds lie at the
heads of tributaries to larger rivers like the Kennebec. Small
dams regulate outflow at the ponds and keep them at predictable levels. But the dams make access for alewives difficult or
impossible. Installation of fishways has become an important
component of the state's restoration program, particularly on
the upper Kennebec. It's not as simple as sending out crews
to install a fish ladder. Maine ponds are choice retirement and
vacation properties. Property owners surrounding such places
form pond associations that dot the state and exercise influence
over the watersheds they feed.

Lawrence and a few colleagues are rendezvousing at Augusta
City Hall to lend support to the pitch for the Worromontogus
Pond Association to approve fish passage at their pond. "It's
a chance to create more habitat," Lawrence explained. "And
you'll get to meet Nate Gray, our alewife restoration specialist.
He's great with these groups, a very dynamic speaker. Plus it's
just important that we're here. A small group can make a tremendous difference."

Inside City Hall, a small auditorium was nearly filled to
capacity, and a PowerPoint presentation had already begun. A
photo of a harbor seal, among the first to be sighted swimming
in the Kennebec above Augusta in two centuries, was projected
onto the screen above the podium. Gray's affable voice issued
forth over the hall, giving an impassioned description of the life
history and ecology of alewives. A football player in college, he
ran this meeting as if it were a science lecture and he was yester-
year's gridiron star turned beloved biology teacher.

"Tell me again, what is the job of the alewife in the ecology
of the rivers of Maine?" he asked.

One woman actually raised her hand before she gave the
answer. "To be eaten?" she asked tentatively.

"Yes! Now who eats the alewives?"

The image on screen advanced to a video clip of a stream

teeming with alewives, some of them being beaten silly by a few larger fish. "Striped bass," Gray announced. "And they've returned with the alewives. And as many of you've seen, we have two kinds of sturgeon, Atlantic and short-nosed, that are also swimming above the old Edwards site for the first time in a while. But there are pieces of the puzzle we're still missing. Atlantic salmon. Shad." The audience seemed mesmerized by the footage as much as they were by Gray's words. But the real selling point of tributary alewife restoration had yet to be made: enlightened self-interest. In addition to their ecological role as the broad base of a freshwater food pyramid, alewives in the local pond serve the same function as zoning boards scheming to keep Walmart out of their towns: they preserve property values. Alewives eat at the bottom of the food chain. They're constantly snacking on phosphorous-laden zooplankton that can cause small, relatively warm bodies of water like Worromontogus Pond to "turn over," becoming an oxygen-deprived, weed-infested swamp. The fish are essentially an organic phosphorous-abatement program, a treatment otherwise accomplished with chemicals that don't come cheap.

After the show, Gray fielded many questions about costs, potential bureaucratic entanglements, timelines, and procedures. When I asked him if he wanted to chat over dinner at a barbeque joint back across the river, my query received the evening's most enthusiastic reply. "God, yes, let's get out of here," he said—and then stayed behind another twenty minutes talking with several more interested pond people.

At dinner, Gray talked about growing up in Gloucester, Massachusetts, during the collapse of the world's most productive fishing grounds. "Here, we're just fishing down the food chain now. You look at the amount of damage versus the amount of oversight, and the damage is always ahead. But the West Coast," Gray observed, "did not learn from the mistakes of the East."

Gray spent the first several years of his biologist career in Alaska. He was in Cordova at the time of the Exxon Valdez spill, came back East a few years after, and has worked the past eighteen years for the Maine Department of Marine Resources. "It might sound like a little bit of hubris, but I think I can make a bigger difference here than I can up there," he said. "In fact, I know I can."

Part of that confidence, Gray admitted, is that Atlantic ecosystems have plummeted so far that there's nowhere to go but up. Of the more than six hundred alewife runs that were once harvested on the Eastern Seaboard, only eighteen remain viable. Atlantic salmon are circling the drain. Shad are struggling. But to Gray, the perks of working in Maine as opposed to the Pacific coast stem from contributing to actual restoration rather than continuing on with a charade Gray sees other biologists often forced to play. "I think in the West we're just going to study salmon until the money runs out or until they're gone, whichever comes first," he explained. "Here, I already know everything I need to know about alewives. Enough anyway to know what it takes to bring them back."

Gray has patterned his life after wise counsel gleaned from Transcendentalist writers who once roamed the territory around the time the first of the region's dams were being built. He lives off the grid, "on a small but failing farm," as he described it, hunts and gathers his own food to the extent it's practical or possible, and understands the work he's doing will pay off in ways that don't necessarily show up in economic or scientific charts. As such, Gray told me he isn't surprised by the ripple effect of the Kennebec's comeback. "People are creeping back up and looking at the river again, and for the first time anyone can remember, it looks pretty good. You know, Maine is a poor state. The phrase 'paper colony' is accurate. It wasn't just the paper mills though. At its peak, when all the mills were running, the river just disappeared under that whole machinery for more than a mile. Look at the architecture of the city. People literally started turning their backs on the river about that time. The term alewife was almost lost out of the common lexicography of Maine," Gray said. "This was a winter source of protein for centuries around here, for Indians and whites alike. Almost all the edible fish are gone, and the Indians went with them. The sad truth is, we're really good at rubbing things out."

Displacing biological wealth and indigenous cultures with the infrastructure of industrialism, Gray contended, has never easily translated into long-term or widespread economic security. The opposite may be just as true. The rise of mills and factories on the Kennebec has some familiar echoes in the fiscal crisis plaguing the country beginning in 2007. When Edwards Dam was completed in 1837, the worst economic disaster in U.S. history to date (until the Great Depression submarined it) was in full swing. The immediate cause of the Panic of 1837 was outgoing president Jackson's disdain for regulation in the form of a central banking system. Jackson didn't just deregulate what we would now call the Fed, he eliminated it all together, revoking the charter for the Second Bank of the United States and withdrawing its government funds. The result was the rapid rise of wildcat and state banks that flooded the country with cheap credit. Venture capitalists quickly borrowed staggering sums of money, mainly to speculate on fire-sale prices for public land. Easy borrowing fueled a madcap inflationary spiral in land values and contributed to a dangerously lopsided trade imbalance favoring imports. States got into the act, borrowing cheap money to finance the frenzied expansion of railroads, canals, and dams, in turn adding more fuel to the speculative fires of private interests on properties adjacent to these projects. The game wasn't limited to a Jacksonian-era version of *Flip This House*. Contestants raced unscrupulously to flip whole forests, fields, swamps, and pastures. Corruption and scandal accompanied a three-fold increase in the value of Maine's timberland in the mid-1830s. Between 1834 and 1836, some forty million acres of public land in the United States were let go into private hands. When the bubble burst, more than a third of the nation's eight hundred banks folded. Real estate values plummeted. Fortunes were lost. Keeping food on the table rather than money in the bank became the primary concern of hundreds of thou-

sands of families. Relief efforts were blackened by food riots
that broke out in several cities. A perfect storm of political circumstances swirled together to rain down calamity on adoles-
cent America. (A hardheaded ideologue as outgoing president,
a flood of criminally cheap credit, an ominous trade imbalance
granting foreign creditors undue influence over the prospect
of economic self-determinism, the incessant fawning over the
prospect of quick fortunes to be made on emergent technologies. Sound familiar?) But the role of the federal government
as a cut-rate broker of easy land deals was undoubtedly a root
cause of the Jacksonian panic attack. The feds had gotten into
the business of furnishing the means for private profit by selling
out resources designated to be held in public trust, a racket it
has since proven difficult to quit.

Given this history, what is it then, I asked Gray, that allows
him to believe he can make a difference? The question caused
Gray to put his fork down. "I don't have a mandate," he answered. "As a biologist, all I have is a mission statement from
my agency. If I had a mandate, everything I want to do would be
done already. What I do have are places like the Sebasticook."
Gray paused for a moment, then leaned a little closer over the
table to emphasize his next point. "I heard that river breathe
for the first time in 170 years in May," he went on. "I could *smell*
the fish coming. I could *feel* them coming. All the other creatures knew they were coming. They were all waiting. Watching.
We passed 1.3 million of them upstream, from a population that
was nonexistent a few years before. I think by the time we're
done, and we get all the population we want coming back, we'll
have around 5 million fish in the Kennebec system. And that's
good, but in all these rivers around the Gulf of Maine, the population was probably 55 million. So there's work to be done.
We're awful good at screwing things up. But we *can* be awful
good at putting things right again."

The biological imperative for doing so, Gray asserted, has
never been greater than now. "If we're lucky to be around as
a species a thousand years from now," Gray said, "historians

will look back on this period, roughly 1800 to maybe 2050, the Industrial Revolution and what's followed, those 250 years, as the make-or-break point of human history. It'll be clear then that natural resources became or should've become part of our national security. Because in the end, natural resources are the only thing that will sustain you. A corporation or your 401(k) does not do the job. We have to believe we can turn this ship around. Because if we can't, there is no salvation. Failure is not an option."

But it seems to me, I offered, that the fervor for dam removal is motivated less by fears over ecological collapse than by hopes for the benefits to local communities. I mention the epiphany of former skeptics like Viles, who'd become one of dam removal's outspoken proponents. Gray conceded the point. "When you see something like a million fish running in the Sebasticook, I think on some level we're reconnecting to half a million years of evolution," Gray explained. "I like to hunt, and I think part of the attraction there is a sense of deciding where your food might come from. In this, at least, I can feel self-determined. There's a sense that the land and water are what really sustain us, that the fate of animals and fish and us is something we all have in common. I think some people feel that connection's more gratifying in some ways than all the trappings of modern life we've come to depend on."

After dinner, we walked across Water Street and down river to a spot Gray told me you can see both short-nosed and Atlantic sturgeon leap acrobatically out of the water. It's just downstream from the old Edwards site and, sure enough, before we get too far into a discussion over "biolitics" (according to Gray, the ungainly mix of politics and biology) one clears the surface by at least a foot. These fish outsize all but the largest salmon and match them in the sheer grace of their cartwheeling over the water. They catapult from the river as if they'd been hooked in the heart. "Short-nose," Gray told me. I could see him shifting gears back into biology-professor mode. "Do you know why they jump?" he asked me. "It's not for food."

Dusk had cast its net on the water. Bridges, abandoned smokestacks, and ghostly church steeples faded to silhouettes. Water and sky seemed more tangible and real than everything else. Some fishermen I know call this the magic hour, ostensibly because the fishing can get really good just before dark. But the metaphysics of conversation by a river at dusk seem to elicit the utterance of ideas that otherwise might not see the light of day. "They're just curious," Gray said after a minute or two of quiet. "They want to take a look around at the world outside the river."

The next afternoon I'm reunited with my family on the Maine coast. We've been invited to a neighborhood lobster feed, courtesy of my brother-in-law, whose clan has vacationed near Boothbay Harbor for many years. A boatload of lobster, corn, and potatoes gets covered with seaweed and steamed over a bed of coals. Children run around on a big green lawn. Older kids fish off a dock. As the sun sinks to the west, a summer feast commences. The evening's repast concludes with several families singing corny old songs together on a big covered porch. At some hour well past dark, full-bellied and sleepy, my own son and daughter and wife pile into the car, and we negotiate the unfamiliar two-lane road back to our rented house near Christmas Bay. On the harrowing drive, I wonder if any of the lobsters we devoured were caught with alewives that Lawrence sold as bait for the traps. It's a goofy thought, driven by the tourist yearning to make connections to places where we're relegated to being mere visitors, gracious guests at best.

The Kennebec is not my place. But I believe what Nate Gray told me about hearing that river breathe.

11

THE HEART OF THE MONSTER

*It's now been a year since the American people went to the polls
and gave me this extraordinary privilege and responsibility.
And part of what accounts for the hope people felt on that day, I think,
was a sense that we had an opportunity to change the way Washington
worked, a chance to make our federal government the servant not of
special interests but of the American people. It was a sense that
we had an opportunity to bring about meaningful change for those
who had, for too long, been excluded from the American Dream. . . .
I understand what it means to be on the outside looking in. I know what
it means to feel ignored and forgotten, and what it means to struggle.
So you will not be forgotten as long as I'm in this White House.*

—PRESIDENT OBAMA AT THE WHITE HOUSE
TRIBAL NATIONS CONFERENCE, November 5, 2009,
informing American Indian leaders of the memorandum
he'd sent directing federal agencies to outline specific
ways they planned to implement or improve "direct,
meaningful consultation" with tribal governments

*It's appalling. But it's also one of the oldest tricks
in colonialism's playbook: divide and conquer.*

—NEZ PERCE REPRESENTATIVE REBECCA MILES,
on the effect on her tribe of Obama's decision to endorse
the salmon policies of the Bush administration

That anyone at all is even discussing removing dams on the
Snake River owes much to the violation of a basic tenet of foreign diplomacy a century and a half ago: never underestimate
the negotiating position of an opponent. In 1855, the federal

government went on a hasty treaty-making tour of the West, hoping to settle land disputes in a hurry, in part to make way for an intercontinental railroad system. Leading the federal family in this case was Washington State territorial governor Isaac Ingalls Stevens. The thirty-five-year-old governor was a monstrously ambitious man back in those days, also serving as a U.S. Indian agent and the head of a western railroad survey. Notwithstanding any potential conflict of interest in these over-lapping duties, Stevens was a successful, if not entirely honest, diplomat. In 1855, he threatened, cajoled, begged, bargained, and stole from Indian tribes from the Northwest Coast to the Upper Missouri, goading these nations into ceding some 64 million acres of land to the U.S. government. The thrust of his diplomatic strategy was simple: take the deal we offer or wind up with nothing, except war with the United States.

Stevens' shtick included promises that the U.S. Army would protect Indians from "bad white men," as he described them, that lands held in reserve by these nations would not be en-croached upon by white people bad or good, and that tradition-al hunting, gathering places, and travel routes through ceded lands would forever remain open to them. "You can rely on all provisions being carried out strictly," he told many Indians.[1]

The Walla Walla Council, site of the negotiations between Stevens and the Snake and upper Columbia River tribes, was an impressive gathering. In addition to Stevens' government con-tingent, some five thousand people representing the Nez Perce, Cayuse, Palouse, Walla Walla, Umatilla, and Yakama nations were on hand.

Stevens and his Oregon sidekick, Joel Palmer, spent the first four days trying to bore the tribes into submission, with Stevens droning on about the rule of law and its ability to affect the behavior of good and bad white people, and Palmer pontifi-cating like a Google executive on the wonders of technology, in this case the endless promise the telegraph and the railroad held for everyone, including Indians. The tribes were not much swayed by epic-length oratory, opting instead to get down to brass tacks.

Stevens' initial proposal was for two reservations, one for the Yakama Nation and one for the Nez Perce. This simply would never do. The governor/Indian agent/railroad surveyor either failed to recognize or chose to ignore the fact that the situation was much more complex than two reservations could possibly account for. Dozens of tribes were represented, and there were clans and bands within tribes that had long enjoyed loose, autonomous, yet permeable territories quite distinct from the Western concept of "nation." In part because of this oversight, Indians in attendance pegged Stevens immediately as a churlish tool.

More than a few Indian representatives were leaning toward tossing both the talk and the government men, who, after all, were guests on Indian lands. Rumors swirled that Stevens and his party would be attacked and killed before the council concluded. To assuage these fears, a Nez Perce representative known to whites as Lawyer moved his camp next door to Stevens. The son of the great Nez Perce Chief Twisted Hair, who welcomed the half-starved, ragtag Lewis and Clark party to Clearwater country only fifty years before, Lawyer had welcomed the first Christian missionaries to the region. He was the first to learn to read and write in his native Sahaptin language. White people liked him. We will never know whether the move next door to Stevens was part of a calculated strategy by the Nez Perce to gain advantage in treaty negotiations, a retreat by Lawyer based on criticism by a few Nez Perce who thought he'd been too willing to parley with the government, or simply a gesture of peace. For better or worse, Lawyer was playing the good soldier on behalf of his people.

The Nez Perce naysayer, the outlaw in Stevens' eyes, was a powerful warrior and medicine man with such a fearsome reputation it was rumored he could kill someone simply by wishing them dead. He was going to arrive fashionably late. The militant yin to Lawyer's peaceable yang, he was riding in fast from the east, making his way home from the buffalo plains.

On the eighteenth day of the Walla Walla Council, Chief Looking Glass interrupted the proceedings in dramatic fashion, and Stevens' initial treaty framework came undone faster than a party dress. Still spry at age seventy, Looking Glass had traveled three hundred miles on horseback from what now is Montana to reach the council grounds in less than a week. He dispensed with the customary pleasantries, delivering his opening speech without dismounting from his horse. His posse was still decorated in war paint from battles on the plains with the Blackfoot. One of Looking Glass' posse hoisted a Blackfoot scalp, dangling from the end of a long stick. "My people, what have you done?" Looking Glass asked. "While I was gone, you have sold my country. I have come home, and there is not left me a place on which I can pitch my lodge." He then basically ordered everyone to go to their rooms: "Go home to your lodges. I will talk to you."

Looking Glass did eventually dismount from his horse, but his invective never wavered much from that hard-charging opening salvo. "Why do you want to separate my children and scatter them all over the country?" he asked members of Stevens' party. "I do not go into your country and scatter your children in every direction." He was vehement about reserving more land: "I marked it bigger. I said yes to the line I marked myself, not to your line." He was suspicious of a treaty that could be negated by the president's refusal to endorse it. He squeezed Stevens for more extravagant promises and provisions to keep whites off reserved lands. Stevens' frustration could not have been more palpable. "We buy their country and pay for it and give most of it back to them again," he hissed in an aside to his aides.

Looking Glass signed the treaty. But his principled stand turned what was supposed to be a signing ceremony the day after his arrival into further heated haggling with Stevens and company. The Nez Perce ceded 5.5 million acres—but kept 7.5 million for themselves. Most important, they, along with neighboring tribes, extracted permanent rights to access ceded lands. In a clause of the treaty that still keeps dozens of lawyers gainfully employed today, the 1855 treaty spells out the terms on which salmon are guaranteed to be a part of any future vision of the Snake River country:

The exclusive right of taking fish in all the streams bordering or running through said reservation is further secured to said Indians; and also the right of taking fish at all usual and accustomed places in common with the citizens of the territory; and of erecting temporary buildings for curing, together with the privilege of hunting, gathering roots and berries, and pasturing their horses and cattle upon open and unclaimed land.

~ ~ ~

One way to gauge how well those promises made by Stevens float on today's currents would be to book a few days' fishing with a guide service dubbed the 1855 River Company. Its proprietor is Levi Carson, a Nez Perce from the Wallowa Band that was forced to flee their achingly lovely namesake valley in Eastern Oregon in 1877. "The Dreamers, Chief Joseph: those guys were my uncles and cousins. Guiding is in my blood," he explained. "Visitors have always come through here, and they realize that everything's pretty much perfect. The grass is waist high. The skies are usually blue. So the Nez Perce have always been the ultimate guides. They gave people fish and meat, tried to assist them on their way. You didn't get killed just for showing up. It was always, 'See you later, and we'll keep the light on for you.' I mean, look around you. How could you be crabby living on this plateau, surrounded by these mountains? It's what makes the Nez Perce Nez Perce."

On a chilly October morning, Carson and I were in the first hour of an endeavor that I was sure would require all the bravery, cunning, patience, and generosity he could muster from the good vibes he gets from his surroundings. He'd foolishly agreed to teach me to cast a spey rod, a two-handed fly rod that, when hefted with the proper grace and finesse, makes throwing eighty feet of line a feat attainable by mere mortals. Problem one: I lacked the proper grace and finesse. At first, I was about as apt a pupil as a fourteen-year-old boy kidnapped from the paintball wars and plopped down in an advanced Tai Chi class, the esoteric pleasures of pushing hands complicated by the addition of a fourteen-foot stick from which the line in question was to be launched. After nearly skewering myself with the fly, some reassessment of technique seemed necessary. "Maybe if we break it down into steps," I suggested.

"No," my teacher flatly refused. "This is like dribbling a basketball. If you think about it too much or try to go in slow motion, you'll just screw yourself." I watched as he effortlessly flicked his line, uncoiling it with ease the length from the pitcher's mound to home plate. "Soft-ly lad. Softly," he said. "You are not driving a railroad stake. Give it a go. And try not to drive your gillie crazy."

The gentlemanly persona melted from Carson's face as he caught a glimpse of another outfitter's drift boat. Guide and clients therein weren't fly fishing but backtrolling through the deepest slot in the river. The technique consisted of rowing upstream, stripping their fly lines out a good distance in front of them, and letting their hooks hunt the river using the boat's resistance to the current, with little effort made on anyone's behalf but the oarsman.

"Are you handicapped?" Carson shouted across the water to them.

"What?" asked the guide, pretending not to hear, looking as if he'd been caught in Sunday school with his hand in his pants.

"Are you running a trip for disabled vets?" Carson asked. "Because if you are, I'd like to help."

Their craft came even with us where we stood on shore, and we had a closer look at their crew. Two clients in the boat appeared stricken with primal fear, the color gone from their faces. Being shouted down by a stocky, braided Nez Perce man in a black cowboy hat and dark aviator sunglasses was not quite the Western experience they'd bargained for.

"Come again?" the guide asked, feigning nonchalance.

"Now don't give me that shit," Carson replied sternly but calmly. "I've talked to you and your boss about this before. The whole idea of having a spey rod in your hand is that you can cast further. But you have to be standing up to do it. Can your clients stand up?"

"Yeah."

"Can they cast? Because if they can't, it's *your job* to be teaching them."

"We can all cast. But we catch a lot more fish this way."

"If you can't catch fish without trolling, go do it in town with the jet boaters. And if you're gonna go that route, at least put away the fly rods and get out a spin rod. You're just *embarrassing yourself* as a guide."

"You fish your way, and I'll fish mine. Plenty of water he—"

"That's just it," Carson interrupted. "You fish your way, and I *can't* fish mine. You'll booger every fish all the way down the river. Ever hear of the idea of fair chase?"

The concept, it struck me, was a resonant one in Carson's past.

Though his knees are bad and by his own admission he should quit smoking, at fifty-four, Carson still sports the physique of a powerful athlete, the rare kind of fireplug that in his youth could choose on a whim whether to run over or around opponents. His fingers are the girth of a garden hose, sprouting from palms the width of a softball. He wears his hair in the manner those uncles did and some Nez Perce men still do, in two tightly woven braids falling in front of the shoulders. He sees guiding as one way to remind outsiders of the terms and conditions agreed to by Stevens and the river tribes in 1855. He takes his role as keeper and guardian of the river seriously. The Nez Perce, Carson pointed out, granted land to whites, not the other way around: "The treaty language says, 'With no conditions attached.' What I do is another way of making sure we're not impeded by the grantees. Stevens lied. And a lot of people's efforts since then have been spent trying to make the lie true."

I'd met Carson the summer before we decided to fish together, and we'd spent hours discussing the finer points of both salmon and Indian politics. For more than a century after that

1855 meeting, we concurred, the federal government willfully ignored its own treaty language, especially that phrase about the "usual and accustomed places."

Stevens' promise to the Nez Perce and other river tribes in 1855 lasted less than a fortnight. A key provision in getting any of the tribes to sign with Stevens was his guarantee that the wheels of the federal bureaucracy would move slowly enough for everyone to adapt to the new boundaries. A delay of two to three years should have passed before any Indian was removed from traditional territory. It was sworn that no white people would harass them. In the meantime, Stevens promised that the feds would build new houses for the headmen of each tribe that signed the treaty and that other provisions, schools, mills, and annuities would materialize in the interim. What happened instead is that Stevens and Palmer hung around the council grounds for a few days after everyone else had left, furiously cranking out a report on their accomplishments, then dispatched an express rider to Olympia, Washington, to distribute the news. Less than two weeks after the treaty was signed, the *Oregon Weekly Times* published an article signed by Palmer and Stevens that proclaimed ceded lands were open to *immediate* settlement by whites, with the caveat that Indians "are secured in the possession of their buildings and implements until removal to the reservation. . . . This notice is published for the benefit of the public. . . . Oregon and Washington papers please copy."

Branches of the Fourth Estate obliged, and all hell broke loose on the plains in the wake of the proclamation. As Alvin Josephy Jr. comments in his authoritative history of the affair, *The Nez Perce Indians and the Opening of the Pacific Northwest:*

> There was no excuse for the premature announcement, and it had its expected result. Almost immediately, land-hunters and prospectors headed across the Cascades, hoping to beat the rush into the new country. If the Walla Walla treaties would not lead to war, this betrayal of the Indians was guaranteed to do so.

Gold strikes in the Snake River country upstream of Lewiston spurred an unruly white horde into the Northwest interior, making half of Stevens' summation of white intent in his council speech come true. There were plenty of bad white men who didn't care much for the concept of law. But a critical shortage of good white men made stemming the Caucasian invasion, along with a wave of armed conflict, a virtual impossibility.

Seven years later, the Nez Perce were hectored by the United States into signing another treaty, ceding six million acres of land. Apparently less interested in justice than bargain-basement value, Superintendent of Indian Affairs C.H. Hale bragged that he'd just increased the real estate of the republic "for a cost not exceeding eight cents an acre." The treaty of 1863, at least in the eyes of the United States, superseded the 1855 treaty, despite the fact that the majority of loosely affiliated clans composing what whites construed as the Nez Perce "nation" either walked out of the negotiations or never showed up. The signature of those who stayed was somehow interpreted as fair proxy for those who didn't. It was a classic bit of colonial strategy on the part of the government: divide and conquer. Agency Indians became known as "treaty" Nez Perce, and those who didn't sign became "non-treaties," a partition that led directly to the bloodiest rift in Nez Perce history, the 1877 flight of several hundred of their men and women from their homeland toward sanctuary in Canada with Sitting Bull's band of Sioux.

By 1890, most Western Indians were sequestered onto reservations, where it quickly became apparent there was no prospect more terrible than that of powerful white people acting on the Christian intent to save them. The Dawes Act, signed into law by Congress in 1887, stipulated that lands held in common by tribes should be divided up among individual families to facilitate their transformation into farmers and their allegiance to the sacred cow of individual property rights. Instead, the law created loopholes in which further transfer of land and wealth from native to white people occurred. Reservation property not squared away into township and range, allotted to individual families, and sold or put to work in some enterprise was considered "surplus" and confiscated by the feds. Indian lands throughout the United States were further reduced by two-thirds or about 90 million acres. Ninety thousand Indians were rendered landless. The social reform prong of the law was an equally destructive force, with Indian children forced to attend Christian churches and schools, where they were made to dress up like fey little suburban Presbyterians and told to forget about being Indian. The reprogramming served only to tear at the remaining threads of a culture already worn threadbare. The Nez Perce fared worse than average, selling or otherwise losing title to all but 90,000 acres of the 750,000 that they managed to hang on to in the 1863 treaty.

If federal intent behind forced removal, followed by a Mao-style, missionary-flavored Cultural Revolution could be described from the government's perspective as the Bad and the Good in this abbreviated tour of federal Indian policy, what came on the heels of World War II furnished the last of this tripartite Western theme, the Ugly. In the 1950s, "Indian termination" was the name given to the idea that if the government could not tailor its legal relationship to the tribes to fit its needs nor reform the Indians themselves, it ought to end its responsibilities to the continent's original inhabitants altogether. It was an unqualified disaster. Though termination was never an imminent threat for the Nez Perce, many neighboring tribes lost their status as sovereign nations, which allowed the federal government to discontinue social, medical, and educational services provided as a condition of treaties with the federal government. Another 1.3 million acres were transferred from Indian possession into private hands. More than thirteen thousand Indians lost their tribal affiliation, a catastrophe that fragmented generational, linguistic, and cultural ties. Though many tribes regained official recognition from the government in the 1980s, the thirty years' termination experience has required another third of a century to recover from. Moreover, the termination era, arguably the low ebb of Indian political rights in the past two centuries, happened to coincide with the years of the Western dam-building frenzy. If Stevens' treaty promise—that "you can count on all provisions being carried out strictly"—had been honored, and Western tribes had enjoyed the free exercise of rights guaranteed to them in treaties negotiated in 1855, few if any of the dams constructed from the Depression to the age of Nixon would have been built.

The preferred alternative to making good on treaty promises abrogated by dam building was simply the logical extension of the negotiating tactic rolled out first by Stevens. Take the deal we offer you or wind up with nothing. The loss of "usual and accustomed places" the respective tribes held most dear, their salmon-fishing sites along the Columbia and Snake and its tributaries, would be reduced to a dollar amount and compensated for with some promise of "mitigation" to be made at the government's convenience—sometimes decades later.[2] The Indians' attempts to get the feds to honor their word has at times become another opportunity for the feds to blackmail Indians into further reductions of treaty rights. Indians from the Colville Reservation in Washington were told in the early 1950s they would be compensated for the portion of their reserved territory inundated when Grand Coulee Dam was built—if they presented a plan for their own termination.

The political undertow of a century of diplomatic, political, and military treachery weighed heavily on the decision of Columbia River tribes to sign off on the construction of The Dalles Dam, which drowned Celilo Falls. Louis McFarland, chairman of the Umatilla Tribe at that time, commented that the Celilo compensation agreement was "comparable to the signing of the treaty of 1855." The Corps, echoing Stevens' high-pressure tactics a century before, sent a letter to tribal leaders within days of receiving the first check from the Treasury to begin plugging the river at The Dalles. The letter stated construction would begin "at the earliest possible date" and that the Corps intended to reach "fair settlement" for the loss of fishing sites at Celilo before then. The Corps hadn't even managed to settle fishing sites lost to Bonneville Dam a generation before. One of the major sites that was supposed to be developed for exclusive tribal use as compensation for "usual and accustomed places" lost to Bonneville now sits underneath the north abutment of The Dalles Dam.

Fair compensation, in the Corps' narrow slide-rule logic, would amount to $3,754.91 per capita. To arrive at this seemingly random figure, the Corps argued that since so many tribes fished at Celilo, it was not one of the "usual and accustomed places" guaranteed in various treaties. Having furnished its own study on the matter, which censored data from its own biologists that found that McNary Dam caused a serious downturn in salmon populations after it was built in 1954, the Corps declared that damming the river yet again would be *beneficial* for fish, since they would be relieved of so much harvest pressure in one spot and wouldn't have to leap those forbidding falls anymore. When the Yakama Nation hired its own fisheries biologist to vet the claims the Corps was making, the Corps, with the help of the Department of Justice, informed the Yakama leadership they would not only reject any requests to share data with Yakama's lone scientist but would refuse to acknowledge his presence in conversation or correspondence, official or otherwise.

This state of affairs was no surprise to the Nez Perce. In 1948, the Nez Perce hired an attorney to contest the construction of McNary Dam by suing the contractor that the government hired for the job. The Bureau of Indian Affairs and the Department of Justice represented the *contractor* in this case, not the Nez Perce. Come hell or high water, the Corps, with the help of the rest of the U.S. government was going to transform the Columbia and Snake. There wasn't a damn thing the Indians could do about it.

The river Indians, nonetheless, kept fighting against The Dalles Dam. The venerable chief of Celilo Falls, Tommy Thompson, who'd refused to be removed to the Warm Springs

Reservation with the rest of his Wyam band, was lead biologist, game warden, and, for most of his tenure, a one-man fish commission at the falls. Thompson never gave an inch when it came to Celilo Falls. He was born in 1855, the year the Walla Walla treaty was signed, and was thoroughly educated about what had been promised then. For more than a decade, he resisted the dam with a zeal equal to any of the world's civil rights heroes, refusing compensation for the loss of the stretch of river that his people had called home since time immemorial. On the day the floodgates were lowered at The Dalles, March 10, 1957, Thompson had finally been removed to the nursing home where he died two years later at the age of 104. "Bring me some more blankets," he called out to his wife on the day the falls disappeared. "I can feel the waters rising. They are covering me up."

Like many Indian fishermen the length of the Columbia, Carson knows something about how Thompson felt that day. While he was finishing his tour of duty for the U.S. Army in the early 1970s, the last of the four Lower Snake River dams, Lower Granite, was completed. "On my way home from the Army, I was driving up the river, and by the time I came down the hill into Lewiston my hands were shaking," he said. "I was like: 'Where are we? Did we just go through the portal somewhere? What the fuck just happened here?' I had never shed a tear in the Army. But by the time I got to the bridge, I cried. I had to pull over. I was trying to smoke a cigarette, but my hands were shaking too much. I was just crying, man. I was just bawling."

Standing in the Clearwater with Carson all morning thirty-five years later, the river just about had me bawling, too, albeit for a considerably lesser reason than a heritable, metaphysical attachment to it. I felt like the real-world analog to the old cartoon cowboy who somehow winds up lashed to a tree by his own miscues with a rope. The morning of casting instruction had been a wash, and we decided on lunch and a drive upriver. Along the way, Carson spotted some fellow Nez Perce fishing guides, and we stopped to say hello. As can be expected where two or more fishermen gather, the talk quickly turned to an exchange of reports on the quality of the angling.

"How's it been?"

"Slow. Don't mind though. Slower days, I get the river to myself. You?"

"I hooked three the other afternoon and evening. Landed none. One big one had me spooled before he broke off."

"You make it up to the Rapid River at all last spring?"

"Oh, you know how that goes. Everyone hangs around the tribal office, waiting for the gossip about whether the fish have showed up. I had a few days off, but by the time I was gonna go everyone else was already there."

"You know how they find out."

"Yeah. Alice. You know, she works there in finance."

"Once she knows, everyone does."

The news Alice evidently shares freely each spring wasn't always so good. The Rapid River, draining the eastern slopes of the Seven Devils Range that forms the east rim of Hell's Canyon, feeds into the Little Salmon River just south of Riggins, Idaho. It has long been one of the tribe's fishing places where traditional dipnetting still takes place every June. About the time Carson got out of the Army, it became the scene of a standoff between white and Indian fishermen that easily could have exploded into the first armed conflict between the two groups since 1877. With many of the "usual and accustomed places" now buried beneath dam-induced reservoirs along the length of the Columbia and Snake, the assumption made on the part of non-Indians, state fish-and-game officers, and some vigilante groups was that Indians fishing "in common with the citizens of the territory," as the treaty language promised, meant that the Nez Perce and the other river tribes would have to fish the way whites did, following the same regulations. Though federal courts had made it clear that this was not the case, the states, particularly Idaho, were slow to recognize these decisions. On the Snake River and its tributaries, these conflicts carried an added measure of intensity. The full consequences of a half century's worth of dam building was quickly driving salmon toward extinction. Some whites, then as now, scapegoated the Nez Perce.

In 1980, Idaho officials announced the closure of the Rapid River salmon season, a ruling they assumed applied to Nez Perce and everybody else. While the Nez Perce fishermen who traditionally plied the waters of the Rapid were devising a strategy to combat the closure, they continued to fish. There were threats, intimidation, and violence—including shots fired at Indian fishermen—increasing the chances of open conflict. Thirty-two members of the Nez Perce nation, including some women and children, announced their intention to exercise their treaty fishing rights in spite of the closure, and rendezvoused at the Rapid one spring morning to do just that. They were greeted with a squadron of Idaho's finest law enforcement professionals, who outnumbered the fishers. Hoping to avoid harm, the Nez Perce joined hands and prayed for guidance. One of the fishermen calmly announced to the cops that "power does not come from guns or numbers but from the convictions of the people" and that Nez Perce ancestors and elders were watching over them. A week later, three hundred more Nez Perce showed up to fish in the old way. In the midst of a national resurgence of Indian pride, Idaho had what was misconstrued as a public relations problem on its hands.

To resolve the conflict between treaty fishing rights guaranteed by the federal government and a federal dam-building program guaranteed to wipe out the fish, the State of Idaho, in sit-down meetings with the Nez Perce, proposed the gift of a "simulated fishery." The tribe could dip their nets and sweep their gaffs in the river provided there were no nets or hooks attached to the poles that make this equipment into working fishing gear. The chairman of the tribe's governing body would be allowed to catch one fish, which would be used for a photo opportunity. All parties would get credit for solving the matter nonviolently. The Nez Perce countered this proposal with a simulated raised middle finger. A slightly better offer came in that afternoon. Nez Perce fishermen would be allowed on the Rapid on weekends. This was a temporary fix, at best. The following summer of 1981, the state tried to close down the Rapid to all fishing again. Eighty Nez Perce fishermen were arrested. Camo-clad SWAT teams cruised Highway 93 through Riggins, and the matter became fodder for the national press that summer.

In 1982, thirty-three separate court cases challenging arrests of Nez Perce men fishing in the "usual and accustomed places" were combined into one decision in Idaho State Court by Judge George Reinhardt. He dismissed all charges, finding that Idaho had not consulted with the tribes on the proposed closure of the Rapid, as the 1855 treaty mandated. At the heart of the matter was an ignorance of the difference between privileges granted by law and rights retained in treaties negotiated between sovereign nations. States' rights do not supersede tribal sovereignty —a fact the feds recognized when they made treaties with various tribes throughout the nation's history.

That Idaho had labored for 125 years under the delusion that the 1855 treaty did not require them to consult with the Nez Perce and other tribes in the management of salmon and rivers implies a lack of clear, consistent direction at the federal level in such matters. But federal behavior where salmon are concerned goes far beyond the pale of benign neglect.

In the same year that Reinhardt issued his judgment against Idaho, fishermen from the Yakama, Umatilla, and Warm Springs nations were facing trumped up federal charges for illegal fishing. The driving political figure behind the case was Slade Gorton (whom we met in chapter 6). As Washington's attorney general back in the 1970s, Gorton had argued passionately against treaty fishing rights. Indians were receiving special privileges based solely on race, Gorton contended, a kind of piscatory affirmative action rather than a provision written into a treaty between sovereigns. A pair of landmark federal court decisions opined that Gorton was wrong. First, in 1969, Judge Robert Belloni affirmed treaty fishing rights, stipulating that Indians could indeed fish whenever and wherever they pleased, within reasonable limits. Moreover, Pacific Northwest Indians, wrote Belloni, were entitled to "a fair and equitable share" of the overall salmon catch, and fisheries managers would have to learn to incorporate this standard into their plans. Profoundly dissatisfied with the decision, the State of Washington sought clarification of the term "a fair and equitable share." In 1974, Judge George Boldt assigned a precise fraction to Belloni's imprecise turn of phrase, figuring it as half of all the salmon swimming in water that fell under the category of "usual and accustomed places." The U.S. Supreme Court upheld Boldt's decision in 1979, and another decision affirmed that the ruling applied to hatchery as well as wild salmon. At least in the courts, the tide had turned in favor of Indian fishing rights. But a courtroom, as the federal family figured, is a long way from the river.

On June 17, 1982, more than a dozen cars full of federal agents, along with a helicopter thundering overhead, descended upon the mouth of the White Salmon River, on the Washington side of the Columbia, directly across from what is now the windsurfing mecca of North America: Hood River, Oregon. The G-men stormed the sleepy fish camp in an invasion they had rehearsed at an abandoned fish hatchery ruined by the eruption of Mount St. Helens. In a raid in which the term overkill would have been an understatement, victorious fish cops came away with no more evidence than a trunk full of legal files belonging to Yakama fisherman David Sohappy. After $350,000 spent and fourteen months of investigation, in which NOAA personnel went undercover to buy salmon from Indians—an act that might have been a misdemeanor had it not been for an amendment sponsored by now–Senator Gorton to make fish-and-game violations a federal offense[3]—seventy-two fishermen from the Yakama, Warm Springs, and Umatilla tribes were arrested. Charged with conspiracy and illegally selling salmon to non-Indians, most were released, but thirteen convictions for illegal fish peddling stuck. (The conspiracy charge, which NOAA had been bucking for all along, was summarily dropped.)

It would be a stretch to ask O. Henry to write fiction dripping with more irony than the feds' confiscation of Sohappy's legal files. Gorton certainly knew Sohappy. A decade before, Sohappy was the plaintiff in the case in which Belloni had issued his landmark ruling in favor of treaty fishing rights. As Washington's attorney in that case, Gorton didn't take kindly to losing that case and took pains to prove he could hold a grudge. Sohappy was sentenced to five years in prison for selling 153 salmon and 25 steelhead, all for less than five thousand dollars. He died in 1991, having spent his golden years fighting to stay out of prison and return to the river he loved.

If anything, the "Salmon Scam" operation proved to be an exercise in hypocrisy. Sohappy wasn't even close to being the kingpin of illicit salmon trafficking. As he was fighting for his freedom, the feds were in hot pursuit of mere *civil fines* against a cadre of non-Indian fishermen who had illegally marketed nearly three hundred tons of salmon.

Sitting on a perch above the Clearwater, Carson and I ate lunch and talked about what had happened on the Rapid specifically and on the big river generally since he was discharged from the Army. "I missed a lot of it," he said. "I fought with booze, with drugs. I got over it, though. The Despise."

"The what?" I asked.

"For reasons you can't explain, you wake up one morning with a heart full of bile. You hate being Indian. Even though you know you might've had the rug pulled out from under you. You just want to be someone else. So you try on new faces. In my time, it was booze, pills, cocaine. Now, too many of our kids want to be black drug dealers." He stopped to crumple a sandwich wrapper. "It's not easy, and sometimes I still carry a heavy heart around," he went on. "It's like some days I have to drag myself out of bed, 'OK Gumby, get out there,' you know? This body I've got, the old Indian superman's just about run out of this one. I just try to keep my go goin' and do what I'm supposed to be doing. But our young ones need to see there are ways to get past the Despise. It doesn't have to be complicated. It doesn't always mean court-ordered counseling. It can be as simple as finding something to sink yourself into."

Carson sunk himself into all the rivers that were part of Nez Perce territory in 1855. A dozen years after picking up a spey rod for the first time, he's become a fixture on the Clearwater, one of its best fishermen. He's become a talented tyer of steelhead flies, a skill that's earned him the admiration of several shops in the area that would like to sell his work. "But it's not for sale," Levi told me. "The only way you get to fish with them is if you come fish with me." Meticulous with the maintenance and care of his gear, he's so mastered the teaching of spey casting that his girlfriend, Lynn, is a better fisher than most men who come to the Clearwater. Yet like many people I've come to know who love the river, he's become outspokenly cynical about the institutions and machinations of politics, including those that govern the Nez Perce nation.

I asked him if anything remotely related to civics ever offers him solace. "I do think what some of us are trying to accomplish will have reverberations not just here but all over the world," he answered. "I look around at this valley and what built it—the trees, the animals, the people—and what I see is that it's all built on salmon DNA. We evolved with them. Our religion, our food, our trade: salmon DNA. We keep the salmon, keep bringing them back, we keep who we are. Self-determined. 'With no conditions attached,' just like the treaty says. But the problem is that there are two lines. There's the line I just told you about, and there's what I call the conundrum line. There are too many people in the conundrum line. Even though that's the line where you catch the Despise. Even with all its dead ends and contradictions. We're still after that shiny piece of the coin. That's the silver lining most folks are after."

Despite his skepticism, Levi has paid attention to and participates in tribal affairs, even though he doesn't place much faith in the process. "I've spoken to the executive committee [the Nez Perce equivalent of a tribal council]," he said. "And my message is always the same. QUIT GIVING AWAY OUR STUFF! But that's the problem with tribal government. Our kids want to act like city thugs, and our government wants to act like white politicians. Colonialism is like that. You ever see the old movie *Invasion of the Body Snatchers?* It's like that. You think you're talking to one of your own, then all of a sudden, bzzt! They've got you. You're one of them."

As we drove back downriver, we passed an oddly symmetrical heap of crumbling basalt on the grassy south bank of the Clearwater. It marks the literal ground of the Nez Perce creation myth. I'd stopped there many times before. A low buck-and-rail fence around the earthly evidence of what the Nez Perce call the Heart of the Monster suggests that those tempted to conquer this fifteen-foot-high peak should think twice. Beyond the well-patronized bathrooms, a less-used, quarter-mile asphalt path leads to a shelter overlooking the Heart of the Monster. The curious traveler is met with appropriate signage and a push-button recording of the creation myth in question. The klaxon that jump-starts the story is always out of order. A box that's supposed to talk about Coyote instead keeps quiet at each push of a button, befitting the canine trickster's character.

Coyote is the embodiment of the comedy that can occur when the angels and devils of a decidedly human nature collide. A hideous, narcissistic, insatiable sexual pervert, incorrigible gambler, frequent victim of his own selfishly cunning and devious schemes, he can also be a benevolent, generous, selfless, freewheeling god, guiding people to good fishing and hunting places, keeping order among the other animals, and negotiating with other gods for the protection and well-being of others.

In the creation myth, all the animals talk and live together. Coyote is busy fishing on the Columbia, building a weir to catch salmon, when the news is relayed to him that a monster is in the process of swallowing up the world. Coyote travels up into the mountains to meet the leviathan head-on. With some long ropes, he tethers himself to the tallest mountains above Hell's Canyon, where the monster inhales him. Far from resigning himself to stew in his adversary's digestive juices, he fights back. Chastising his fellow belly-bound prisoners for not taking more decisive action to at least rumble the belly of their oppressor, he produces a flint and starts a fire. Impatient with the slow progress of smoking his way out, he finds a knife and begins cutting out the heart of the monster. As smoke pours from the suddenly vulnerable beast's eyes and ears, the monster is finally forced to open his mouth, and all the animals make a break for it. (For comic relief, some animals escape out other orifices.) Victorious as Saint George, Coyote cuts up the beast, flinging pieces of the corpse over the mountains. Wherever pieces of this mythic roadkill land, a tribe of people sprouts. Fox reminds Coyote after the frenzy of slicing and dicing, chopping and tossing, that no people had yet been made on the ground where Fox and Coyote stood. The monster had already been distributed far and wide. Coyote washed monster blood from his hands, and from this the Nez Perce grew. "They will be small in number but powerful," declared Coyote.

Back down river, with trepidation unbefitting any mythic trickster-hero, I rigged up to have another try at learning to spey cast. "So do yourself a favor," Carson offered. "I know you've got a lot going on *up here*," he said, tapping the side of his head. "Put all that aside this afternoon. It's just this rod, this stretch of river, and you. Nothing else. That's why we do this. Cheaper than psychotherapy."

The impromptu motivational speech helped. I had a passable imitation of a cast down by mid-afternoon. We kept at it till dark, and by then I was looking forward to a full shift of slinging line the next day.

I crashed at Carson's place in Lapwai that night. The house he shared with Lynn was compact, and though it was filled with the signs of the bustle of running a guide service, as well as keeping tabs on children and grandchildren, it was neat and orderly. I asked Lynn about the landscape paintings that adorned the walls of their home, including one of a figure that looked suspiciously like her fly fishing in the Clearwater. "Levi did all those," she told me proudly. "He's just shy about it. I told him, and other people have said to him, he could paint for a living. But they're like the steelhead flies. Not for sale."

At the kitchen table, Carson guided me through his latest artistic fly creations, taking care to never touch the hackle or wings and restoring symmetry to a few specimens that had gotten out of balance in the river's current or a steelhead's mouth.

Fly rods were leaned up near the front door, ready for morning. Watching over them, hung high on the entry hall wall, was an ornate ceremonial headdress adorned with many feathers. At the kitchen table, the plumage of other birds was stored in plastic bags around a fly-tying vise. In Carson's world, some birds are granted reincarnation as guardians of cultural traditions, others as hunters of fish. This struck me as a Nez Perce twist on a Western literary classic. As in the Maclean clan depicted in *A River Runs Through It*, in the Carson household there appeared to be no clear line between religion and fly fishing. But the modus operandi didn't have much to do with escape from the sodden strictures of Scots Presbyterians. What began as an aristocratic, occidental pastime had become a quiet little ceremony in Carson's nimble fingers, a private offensive against the Despise.

I studied a dozen or so of what struck me as his most beautiful flies a little while longer. The more flamboyant red and orange winter streamers appeared as miniature Molotov cocktails exploding from the eye of the hook or a cool imitation of a Zippo accidently cranked to eyebrow-scorching heights. Weapons custom-made to burn the heart out of a monster.

~ ~ ~

In 1999, the four treaty river tribes (Yakama, Warm Springs, Umatilla, and Nez Perce) issued a report from their own fish and wildlife agency, the Columbia River Inter-Tribal Fish Commission (CRITFC), addressed to the Corps. It outlined their position that the four lower Snake River dams should be removed: "From treaty times to the present, non-Indians have taken most treaty protected assets of value from the tribes, particularly their land, waters and salmon . . . by decreasing the salmon productivity of the Snake River system, [the dams] have effectively transferred wealth from the tribes to non-Indians." Thus the campaign to remove the dams gained a powerful ally. Since the days of the Rapid River standoff, CRITFC, or one of its member tribes, has driven many of the legal and political victories that have kept some species of salmon swimming in the region's rivers.

After three decades of losing ground to tribal fishing interests, and with their backs to the wall in the ongoing saga of the bi-op before Redden's court, the feds decided it was time to revisit a strategy from the past. In July 2007, I met with one veteran of the salmon wars who'd worked closely with the bi-op case for its duration. Preferring to speak anonymously, he predicted with precision what the BPA was up to: "They're losing in court, so they're going to try to buy off all the plaintiffs. And you can bet they'll be reminding everyone who funds their projects." A year later, his tip proved true.

The feds offered to spend $980 million over ten years to expand existing tribal fisheries projects, as well as to initiate new ones. But the proposed agreement harbored a familiar ghost. In order to cash the checks, the CRITFC tribes would have to swear an oath of loyalty to the very thing they'd spent years battling in court. After twenty years of fighting tooth and nail against the dammed river system, the tribes would be obliged to switch sides in Redden's court and be seated with the defendants rather than the plaintiffs in the case. Previous testimony made by the tribes on the plaintiffs' behalf would be rescinded. Finally, the breaching of the four Lower Snake dams would be swept under the rug. The tribes could not even mention it publicly until 2017.

Like the litany of treaties before it, the Columbia River Fish Accords, as this deal came to be known, are predicated on sweetening a sour deal with a dose of cash and the promise of remedial action to come later. Worse, as the river tribes have

taken great pains to document, the BPA has a habit of under-
valuing the cost of its salmon recovery commitments. With the
Fish Accords, the BPA would finally agree to fund the program
at an adequate level, but only if the tribes endorsed the idea
that their own tributary habitat restoration—without any big
changes to status quo dam operations—was all that was needed.

Details of the accords reveal that little more than half, or
$540 million, of the nearly $1 billion goes to fund new projects. The balance goes to "guarantee" funding for projects
already underway. But as the Nez Perce, along with the remaining plaintiffs pointed out, funding for these endeavors should
have been guaranteed anyway. Before the Fish Accords were
signed, the feds had written up many of these ongoing projects
as mitigation actions in the bi-op. To claim that work like habitat restoration improves the odds for salmon survival, ESA law
mandates such actions "reasonably certain to occur." The first
thing that disappears after the money is gone is certainty, especially for any kind of public works project.

The subtext of the dickering around the Fish Accords, then,
was a throwback to 1855 and Stevens: take what we offer you
or wind up with nothing. It was not entirely a surprise, either,
when some of the CRITFC tribes signed on to the deal. In
tribal economies where unemployment runs from 25 to 45 percent, rising as high as 80 percent in the winter, keeping poverty
at bay devours a lot of tribal resources. Median income in Lapwai, for example, hovers just over $10,000 a year. Many of the
employees of tribal fisheries programs come from families who
used to fish the river for a living. One tribal fisheries manager
described the federal offer sheet as "blackmail." "We either take
the buyout," he said, "or cut jobs and programs."

The Yakama Nation stands to gain the most from the Fish
Accords, with some $330 million slated to help salmon and other
aquatic species return to the many rivers that once defined their
"usual and accustomed places." But some tribal members saw a
strand of extortion in the fine print. When a vote on the matter
came before the Yakama General Council, in April 2008, they

initially turned down the prospect, voting 39–37 against it, with 69 voters abstaining, saying they didn't have enough information. Tribal elder Delores George told the *Yakima Herald-Republic*, "If you give up something for that amount of money, you're always going to lose something vital." But in an unusual procedure, tribal chairman Ralph Sampson Jr., disappointed by the refusal of the BPA's offer, called for a revote the very next night. He somehow drummed up 152 more yes votes than he had only twenty-four hours before. One reason for the miraculous change of heart was that the BPA had graciously offered to send a few representatives to speak to the council on the perks contained in the offer. Whites are not usually allowed at council meetings.

Through the Wallula Gap and into Snake River country, however, the BPA's lobbying had fallen on deaf ears. The Nez Perce had already answered the BPA's lucrative overture. Their answer would echo Carson's policy regarding his steelhead flies: not for sale.

The Nez Perce point person on salmon issues since 2004 has been Rebecca Miles, a young single mother thrust into the spotlight of the traditionally male-dominated worlds of tribal and salmon politics. When she spoke at negotiations on the Fish Accords, her voice carried a familiar eloquent indignation. The implication was made by someone on the feds' side of the table that the Nez Perce were holding out for more money. Miles delivered an earful, reminding the opposition it was they who'd been holding tribal fish projects for ransom. "You'll cut my legs off, then offer to sell them back to me only if I come over to your side," she snapped.

After the three other tribes had signed on to the accords, the BPA's Steve Wright seemed bewildered by what he perceived as the Nez Perces' delay. "C'mon you guys," Wright said, directing his attention at Miles and the Nez Perce contingent. "I thought this was going to be a celebration today." As in 1855 with I. I. Stevens, what was supposed to be a signing ceremony turned into another round of heated negotiating. But this time, there was nothing to bargain for. Wright thought maybe Miles had just gotten up on the wrong side of the bed and made a trip all the way out to Idaho a few days later to address the members of the Nez Perce Tribal Executive Committee in person. They assured him there was nothing wrong with his hearing when he received Miles' reply.

"This was really a community decision," Miles told me in the cafeteria of the tribal community center. "Many, many people spoke their minds. So that when I took the message to the negotiating table, it wasn't just me talking. What it came down to was that we are the tribe that has the most to lose by not talking about breaching the Snake River dams. We respect the decisions the other tribes made. But we feel like all the options had to be on the table. We've advocated dam breaching along with the other tribes for a long time now. Like the other tribes downstream, salmon are a huge part of our culture and our religion and economy. But for us, our salmon have to deal with those dams before we can fish for them. Getting some kind of major change done with the dams is a good thing for us to fight for. We have a judge and a new administration that seem willing to implement change. So to give that fight up right now was something almost everyone felt was the wrong thing to do."

It's not as if the modern Nez Perce nation eschews the art of compromise. In 2005, the tribe signed a settlement agreement with the State of Idaho over water in the Snake River Basin. Having lost all but a sliver of the land they carved out for themselves in the Stevens treaty, the tribe now faced the prospect of losing claim to a significant portion of its water. Controversy over tribal water rights has spawned a whole new branch of legal scholarship. One source of complexity stems from a 1908 U.S. Supreme Court decision in *Winters v. United States*, which decreed that if the right of prior appropriation—"first in time, first in right"—was indeed the governing principle for water allocation across the nation, then tribes have a reserved water right that dates from time immemorial. Moreover, due to treaty provisions that guarantee off-reservation hunting and fishing rights, tribes also hold an implied water right that extended beyond reservation borders. Upstream users could not do anything—build dams, diversions, or reservoirs—that would compromise tribal water rights. Like many court rulings favorable to tribal interests before and after, the *Winters* decision for most of a century was eclipsed by the implied Western sentiment embodied in the right of prior appropriation itself: possession is nine-tenths of the law. By the time some extremely slow-turning wheels of justice caught up with *Winters* in the 1980s, water rights in Western states were being "adjudicated," meaning states were taking on a lengthy, expensive, and complex process to finally figure out what water belongs to whom. This was the perfect environment for an all-too-familiar brand of cynical politics to sprout.

Employing some especially contorted legal gymnastics, a special Idaho water-rights court determined the Nez Perce had no off-reservation water rights. Though the *Winters* case and a convincing caseload of follow-up decisions had made it clear that tribal water rights were not limited to the boundaries of tribal property, Judge Barry Wood was easily convinced these legal precedents didn't apply in his court. In Wood's opinion, treaties were made to facilitate white settlement, not protect tribes. The implied water rights of the Nez Perce, if granted, would pirate resources from latter-day irrigators and developers as they patched up any leaks in the ship of manifest destiny just as it was docking at its final destination. It was later revealed that Wood owned property and accompanying water rights that would be affected by the outcome of his decision and probably should have recused himself. That he opted for less than full disclosure could easily have been interpreted as a violation of Idaho's Code of Judicial Conduct.

The Nez Perce gritted their teeth, filed an appeal to the Idaho Supreme Court, and lost. The next move was either another appeal to the U.S. Supreme Court or an agreement to sign a compromise agreement with the State of Idaho. The current lineup at the highest court in the land has taken a rather low opinion of Indian rights. Was Stevens haunting the tribe one more time? "It's still hard for me to tell you how frustrating that whole process was," Miles told me. "In Indian country, it's not so much your sex but your age that will make other people skeptical of your qualifications. Experience, patience, wisdom: things that come with age are highly respected. And here I am, this thirty-something on the executive committee. And I can tell you it was no fun to have to report back on what was happening with our water. I'd been granted a level of trust that I've worked hard to keep. I know we have limited resources. I know we have to compromise. Look at our history. If we hadn't, everything would have just been taken from us. Maybe even they just would have killed us all. Even now, I know we have to pick our battles, limit them to the ones where we have a decent chance to win. And it was hard to come back to them and say that we were losing this one big time."

Loss cultivates wisdom more quickly than easy gain. Miles was soon making multiple trips a month to Portland and BPA as well as CRITFC headquarters to represent the Nez Perce in their downstream interests. It wasn't long before the concept of the Fish Accords was floated to the CRITFC tribes, and Miles realized the politics of salmon made the politics of trumping tribal water rights look like a school board meeting. "I was naive at first," she recalled. "Both Bob Lohn and Steve Wright are total schmoozers. I mean, these guys will sidle up to you and ask you about your kids and show you pictures of their families on their BlackBerrys. They'll make you feel like they're on your side. But after things fell apart, it became so clear that they were acting that way only because the tribes had something they wanted. Steve Wright used to call me just to talk. But I haven't heard from him since the deal with the other tribes was signed. The worst thing is, this is not the beginning of a new era of cooperation the way that BPA has managed to package it to the press. There's a lot of dissension out there, even among the tribes who signed. My mother gets on these chat sites, and she talks with people from the Yakama side. They say, 'We sold our most valuable asset, the right to talk.'"

All of that had occurred before the height of summer 2009, when hopes were still riding high that the Obama administration and a sympathetic federal judge might be enough to validate the Nez Perce decision not to sign the Fish Accords. Then it became clear that the federal family had been successful in their bid to sell the Obama administration on the Bush-era biop. "I'm appalled," Miles told me in late 2009. "The science on the issue is silent, and the politics have taken a front seat. When I met with Dr. Lubchenco last May, I really felt like she understood this issue, and even though she wasn't allowed to talk about specifics, I was very optimistic. She basically said, 'We get this.' I trusted her. I hope she's just going through some of what I did early on, when I was new to the salmon issue. It takes some time to figure out the situation, how persuasive that side can be."

To Carson, there's not much point in waiting for bureaucrats to see the light. "Everything's taken," he said. "Our language, our land, our water. And even now, they can't get what they want. So they try a different approach. 'Oh hey Indian, haven't you heard? Now we're going to *share* everything. First give us your half.' Meanwhile, there's a dusty old box in some basement room or some attic that has all the provisions of that 1855 treaty spelled out. It says, 'With no conditions attached.' Time to dig up that box. Take a good look at what's in there. Because if we don't, they're just going to keep coming for our stuff. Until there's nothing."

Dawn broke clear on our last day of fishing. With frost on the ground, I shivered miserably as penance for forgetting my winter coat until the sun cleared the tawny hills above the canyon. About the time things finally warmed up enough for my testicles to make the descent from near my Adam's apple south, I felt the first nudge of a steelhead at the end of my line. I saw a pyramid-shaped head and a dorsal fin, felt the twitch and twitter at the end of my rod, and, maybe only because my blood was still jelled, resisted the temptation to jerk the rod in what I was sure would be a futile attempt to set the hook.

"Jesus, lad. You're driving your gillie crazy," Carson observed. "Why didn't you try to set the hook?"

"Too cold, I guess."

"It was a weird take. I'll give you that. Maybe try a different fly."

A mere five hours later, after happily forgetting about everything except the rod and river, soaking up the heat on one of the last warm afternoons of the year, the end of my rod finally did not twitch or twitter but was bent brutally toward the water. Against all odds, I'd hooked my first Clearwater steelhead on a spey rod. It was a short fight. Before I could even think about getting a crank on the reel, all my flyline and a good fraction of the backing had been won by the fish. When I offered some resistance in this lopsided tug of war, the surface of the river shattered into silver splinters. Two spectacular displays of acrobatics followed, backlit by the autumn sun. Much to my surprise, I was still attached to the steelhead. I could feel her rolling and shaking her head in a calm, deep pool with a wide, shallow bank. A perfect spot, I thought, getting way ahead of myself, to land a fish. Then she was off.

After cussing myself out for a minute or two, I saw Carson approaching from upstream. I was surprised to see how far the fish had taken me. He slapped me five and gave me a quick embrace. I affirmed my stumblebum status by slapping him too hard on a tender shoulder that had been operated on a few months before. His eyes lit up like he'd been electrocuted. "Soft-ly, lad, softly," I joked, the humor apparently enough to keep my gillie from braining me with a rock.

"I was watching you cast for probably twenty minutes before you hooked up," he said. "It's looking pretty good for you. Yesterday morning, I didn't think I'd ever be saying that." His face held the smug satisfaction of a teacher finally getting through to the worst shithead in class.

We fished until after sunset, rowing in the dark to the boat ramp near the tribal casino. On the float out, talk turned once more to the uncertainties of the future river. "The time has come to take those Snake dams out," Carson said. "The recovery would get the next few generations a long way toward fulfilling some of the promises made in the treaty." A smile hit his face like a wave. "But wouldn't it be interesting," he posed, the artist's subversive imagination at work, "if they let everyone loose on the dams right before blowing them out. Just to paint or write whatever was on their minds. I bet you'd get some real interesting messages on that blank canvas."

Under the influence of the drone of semitrucks and the lights of the casino, those nagging practical political considerations, the evermore sacrificial compromises made to perpetuate a civilization too often unjust, got the best of me. What about the wisdom of picking your battles, staying alive to fight another day, wise husbandry of scarce resources, including money and political clout?

The last word was his. Carson tapped me softly on my shoulder, the final gesture in two days' worth of subtle redirections. "It's true that we can't win every fight," he said. "But we *cannot* back away from all the fights we cannot win."

We parted ways but not before he and Lynn gave me some smoked chinook. The package did not make it home. Heading home west in the dark, I ripped open the plastic seal, ate heartily, and thought a little more about the exploits of that Nez Perce joker who beat the beast that tried to eat the world.

When Coyote was still on the earth, the valley was covered by a big lake. Coyote could see that the salmon would come up from the ocean if he would make a hole through the mountains. So Coyote with his powers went down and dug a hole through the mountains. The water went through the hole and on to the ocean. The water in the big lake up here was drained, and the water coming out of it made the Columbia River. Then the salmon came up the river to this part of the country. His people after that always had plenty to eat.[4]

He's digging to free the river again.

THE RIVER WHY NOT

It is by seeing things as better than they are
that one arrives at making them better.

—KENNETH GRAHAME

The plane could not fly in a rare two-day summer storm, so if we wanted to see Buzzkill Creek, we'd have to walk there. We'd already burned half a Friday off figuring this out, sipping coffee and staring out the window, trying to muster the courage to hike in the rain. "It's forty-five miles. We have forty-eight hours. That's less than one mile per hour," my friend dead-panned. "It'll be easy. It's just walking."

Taking a decidedly anti-alpine start, by three o'clock in the afternoon we were slogging in a steady deluge up one of the main veins draining the Idaho batholith, a roughly two-hundred-mile-long, hundred-mile-wide heave of granite that harbors the largest wilderness area in the Lower 48. It also contains dozens of rivers and hundreds of creeks once inhabited by millions of chinook salmon. Access to the biggest and best remaining sanctuary for Pacific salmon outside Alaska was cut off with the construction of those four lower Snake dams. I was after some vestige of that former plenitude. The mission was really a simple one. I wanted to find some chinook that had survived the gauntlet of dams to return to their home streams a thousand miles from the ocean.

By five o'clock, we were taking advantage of a brief respite from the downpour on a broad riverside beach to boil water for freeze-dried cuisine. The dripping pine and cedar and the wet sand lent the feel of an April hike in the Coast Range rather that in the August-arid Idaho mountains. "Why again are we doing this?" my friend inquired.

"To see some fish?"

"Seen 'em before."

"For fun?"

"If we wanted to have fun, we would have floated *down* the river. Walking up it seems a lot like work."

"The chore of a big hike," I observed, "is all a matter of degree. This walk seems pretty epic by weekend-warrior standards. But the people that lived up here before probably didn't think much of it."

"Yeah, well, they probably didn't leave for a weekend on the Buzzkill at three in the afternoon on Friday either."

"You might still have it your way. We might be building that log raft to get back down here if the weather doesn't clear by Sunday. If it's still raining like this, the plane won't make it to fly us out."

Another inspiration for the gamble with the weather was a keen interest in the Buzzkill's rich and ironic past. Before this area was a part of a vast, modern wilderness, it was the cradle of a small but durable civilization. All Nez Perce people, according to one rendition of their history, can trace their origins to the site of a former village there. It's still eighty miles from any paved road and twice that distance from the nearest full-service town. Gleaning from an archeologist's interpretation of the site, you can imagine why the Nez Perce would choose the Buzzkill as a place to begin. The high mountain peaks, as much as a mile above the west-facing valley, suck moisture-laden winter storms to their flanks, creating a classic rain shadow lower down. Mean winter temperatures hover just above freezing. The Buzzkill gets only about half the annual rain and snowfall for the region. It is the warmest place on average for fifty miles in any given direction. The miraculously milder climate, plentiful game, close access to bison-hunting territory to the east, and a stream teeming with chinook salmon provided a secure supply of food. For unknown reasons, the village was abandoned about the same time that Lewis and Clark stumbled into Snake River country, but the Nez Perce continued to make annual pilgrimages to the bison plains via the Buzzkill Basin for more than a century after that. This place's incarnation as untrammeled wilderness, "where man is only a visitor," is a recent designation. For a long succession in a clan of mountain Nez Perce, it was home.

Even after no permanent village remained here, the route was used for another 120 years to access bison-hunting grounds to the east. Walking up the river, through the mountains, and onto the plains and back must have been a journey that promoted patience, endurance, and good humor, along with the idea that getting home always requires a healthy dose of moving on.

In the spirit of that tradition, we proceeded on long into the night, stopping only to sleep in a damp tangle of driftwood for a few hours. Getting in a little brisk walking before breakfast the next day, we stopped to study some fist-sized tracks fairly fresh in the mud.

"Dog or cat?" I asked my friend.

"I've seen a cougar before. But I've never seen a wolf. So," my friend concluded with his accustomed unassailable powers of deductive reasoning, "let's say it's a wolf." We followed those tracks intermittently for a good part of the day.

Just before dark on Saturday, we finally arrived at the Buzz-kill. More leaping fish than could be reasonably counted made it look like the creek was raining silver trinkets from its cobbled bottom up toward the faded blue evening sky. We boiled more water for more freeze-dried victuals, soaked our feet in the cool creek, unrolled our bags, and drifted off to sleep. A stone's throw away lay another clue to the destination of whatever critter was laying down those burly tracks in front of us. Tucked into a thicket of scrubby pines was a boneyard littered with leftover deer parts. More than one unfortunate ungulate had been butchered and, from the looks of it, shared among several eager diners.

East of here in Yellowstone country, the return of wolves has had the serendipitous effect of improving the fishing. Deer and elk, having enjoyed a half-century's hiatus from the task of hiding themselves from *canis lupus*, have quickly gotten reacquainted with the precaution of not eating the flora that provides cover along a stream bank. The resulting recovery of this riparian zone has translated to better habitat for Westslope cutthroat trout. Biologists call these ecosystem-wide reverberations prompted by the absence or presence of a single species a "trophic cascade."

Wolves are precipitating a food-chain fountain of regeneration in the same way salmon could be performing the same good work over a much larger stretch of country to the west: an abundance of good habitat. More than five thousand linear miles of salmon-friendly streams lie in the three million acres of the Salmon-Selway-Bitterroot wilderness complex. With the looming specter of climate change, this generous refuge could turn out to signify long foresight on behalf of those who fought hard for it almost half a century ago. Freshwater ecologists expect a warmer world hereabouts to raise stream temperatures, bring earlier snowmelt, and reduce summer flows. Usable habitat for chinook and steelhead in lower-elevation basins may be cut in half. Higher elevations may see less catastrophic habitat degradation and offer the last, best hope for salmon to survive in a rapidly warming world. And though the Snake River Basin has lost almost all of its once-prolific salmon runs and most of its pristine habitat, its wilderness tributaries flow protected within this big country, with more than fifty genetically unique wild populations of salmon and steelhead eking out a living around the high wild reserves in the Idaho and Eastern Oregon mountains. It seems it will be imperative to conserve these valuable wild salmon and steelhead of the subalpine reaches of these rivers. Keep whatever salmon have survived thus far soaking in the cold batholith, and there's at least an outside chance they may make it to the next merciful planetary cooling trend. But the dire emergency they currently face can't yet be pinned to global warming.

Back at the Buzzkill, the alarm went off at quarter to four. The eerie phosphorescent light of a summer moonlit night made the trail appear like a scene out of some wide-awake dreamscape. We eschewed breakfast in favor of more walking. Without packs on our backs or food in our bellies, we felt like we could almost fly. As the first rosy pink cheek of dawn appeared on the eastern horizon, we came to a place where the Buzzkill, which runs through a fairly narrow canyon for several miles upstream from its mouth, abruptly opens into a broader valley. Marking this sudden transition from canyon to meadow is a deep emerald pool, guarded on one side by a high granite cliff that drops straight into the drink. Perched eighty feet above the water on the edge of this precipice, we could peer down into the depths and survey for chinook. The water was opaque in the low morning light. We waited for the sun to climb high enough to light the water. The shallower edges of the pool lit first. Just then I caught a fleeting glimpse of a single, small, forlorn chinook on the far side of the creek. The sunlight had exposed her cover beside a car-sized boulder. Perturbed at the intrusion, she wriggled into the depths of the pool like a shy, sullen teenager. We stayed until the sun laid the pool bare. One salmon was all the Buzzkill's near-perfect spawning water had received late in that year.

As Noah knew, for the purposes of propagating a species, one is a meaningless upgrade from none. In the manifold tributaries of the big rivers that drain the batholith, the annual pulse of millions of salmon has become the faintest of heartbeats. Extinction is an event not on the horizon but of the season on the gin-clear water of the Idaho wilderness. In poor water years, relatively small drainages like the Buzzkill, once strongholds for millions of chinook, don't see any fish. In years of decent run-off, which are forecast to become fewer and further between, a few dozen may make it back. The government's scientists have pointed to the endgame for this grave trend. According to a 2006 report, there's a better-than-even chance none of the native Idaho, Eastern Oregon, or Washington stocks of wild salmon will make it to the next century. Unless drastic changes are made, the report put the risk of extinction for salmon in these pristine rivers at 55 to 100 percent.

The best available science indicates that the eight dams between the wilderness and the estuary are the primary cause of this risk. A 2007 study, written by Idaho Fish and Game and U.S. Fish and Wildlife Service biologists, found that Snake River chinook survive at just one-quarter to one-third the rate of their downriver counterparts, despite equal or better spawning habitat. The authors concluded, "Recovery strategies must reduce both direct and delayed components of hydrosystem mortality, regardless of the actions taken at other life stages." Habitat, harvest, and hatchery effects are the varicose veins in a flat-lining heart patient. Hydropower is a blocked coronary artery. Ironically, what can be glimpsed in some of the wildest, cleanest, coolest waters remaining in the country is the extinction power of the eight dams downstream. Wilderness as an isolated sanctuary—precious, mysterious, and awe-inspiring as it is—cannot alone provide what salmon need to survive.

~ ~ ~

The salmon's plight may be clearer when viewed from the window of a plane. In the nick of time, we did catch that Sunday ride out, and as our puddle jumper buzzed out of the Buzzkill Basin—waves of mountains starboard, the tawny plains of the Palouse to our port side—it became a little easier to conceive of Idaho as merely the eastern edge of the salmon's habitat, a range that stretches to the end of the continent and across the Pacific to the Asian coast. To restore populations flirting with extinction, to protect what wild stocks remain, they'll need nothing less than a healthy, intact Pacific Rim ecosystem: the mountains, forests, rivers, ocean, and atmosphere performing their usual good work well enough to keep the epic saga of salmon evolution in play. If these creatures are to continue on a ride that began with the melting of the Ice Age glaciers, it will be necessary to cultivate a quality some Nez Perce found in their long walk through the mountains: a more generous, balanced, and complex conception of places that collectively point the way home. The scale of such an unprecedented, well-orchestrated international effort seems well beyond the anemic possibilities offered in the dysfunctional arrangements of Western politics.

This sad reality alone should not be cause for despair. A fair reading of history can be taken following the premise that revolution in the halls of power is merely a victory lap. The contest begins long before that, when two or more people act on an inspiration that reliably derives its power far from any congressional or government offices. The people will speak from unexpected places.

Down on the docks in Astoria, Oregon, the last surviving commercial salmon cannery in town belongs to Steve Fick. I first met him a long way from the wide mouth of the big river. He'd come to Spokane to speak with some of the locals to map whatever territory they might have in common in envisioning a future for the Snake River. We introduced ourselves in the bar at a downtown hotel. I'd only asked him one or two questions when his phone rang. "You're on my shit list, buddy," he said into the receiver, turning to wink at me. Fick took a solid ten minutes gently chewing out his teenage son: no, you can't have the car whenever you want; yes, you had better start shaping up at school; there will be hell to pay if you can't follow through on the following items; I love you; and things had better be on track by the time I get home. It was a judicious deployment of clear expectations, subtle yet authoritative moral support, and good humor—a combination that has obviously served Fick well in both family and business matters. He hung up the phone, stirred a vodka and cranberry juice with the cocktail straw, and, without a prompt, picked up answering my question from ten minutes ago right where he'd left off.

"As the salmon runs go, and the work that goes with them goes, so goes the community," he told me. "A social worker I know in Clatsop County [where lies Astoria] wrote a paper that pegs rates of teen crime, suicides, and the like to the health of the salmon runs." I asked him if he thought the phenomenon was similar to the rash of despair that plagues some native communities when the sense that all that grounds their identity seems already gone or fast disappearing. "Well you'd have to read the paper for the answer to that," he said. "But it makes sense to me. It's real. I've seen it."

Like many Astoria fishermen, Fick now operates in Alaska as well as Astoria to make ends meet. But he's Astoria born and bred, and quite happy to call the quiet coastal town at the mouth of the big river home. He makes a decent living running a small, flexible canning operation that puts out a product a cut above the standard tin of tuna. But he's troubled by patterns of the past and prospects for the future. "People don't realize what we've lost here, and what, over the past century, I think can be fairly said was taken from us," he went on. "In the 1930s, the richest town in the United States per capita was right across the river on the Washington side. It's called Chinook [after the salmon]. Even now, everything over here, the ball fields and the museum, the swimming pool, the churches and restaurants and the hospital, the college, a good portion of the infrastructure of the town itself—all that was generated by salmon dollars. The wealth of the river is just going elsewhere now."

Fick admitted the history of commercial fishing is checkered with excesses. "We made mistakes," he confessed. "But since the dams went in, the mistakes have not been ours alone. We have become conservationists out of necessity. The farmers and the power interests keep asking for and getting more and more. Fishing communities have been forced to get by on less and less. Now the prescription to save Astoria seems to be predicated on selling hotel rooms and restaurant meals to weekenders from Seattle and Portland. The kids are supposed to get a good education, then move away to good jobs in the cities. That's just not a fair trade. That's not a diversifying rural economy. In a sense, I am in the same business the farmer is, which is putting decent food on people's tables. And what we have in common is that it's got to be one of the most important kinds of work there is. And it's become less and less a viable way to make a living."

One of those farmers Fick met on a previous visit to the Palouse was Bryan Jones, who grows wheat on a section of land homesteaded by his great grandfather near Dusty, Washington. On a breezy autumn afternoon, I met Jones at the Central Ferry bridge on the Snake, just down the hill from his farm. Jones had two kayaks in the back of his pickup. With an upstream wind, we planned on paddling upriver to a peach orchard near the site of the drowned river town of Penawawa.

The history of the orchard strikes me as yet another lesson in the art of the possible, another path of resistance that defies the shallow narratives of conventional politics. The family that owned a larger farm at the site of the present peach orchard was bought out by the Corps in the early 1960s. Most of their farm, they were informed, would soon be under several feet of water. Having no other realistic alternative, this family accepted the Corps' payment, bought land elsewhere, and, over the ensuing generation, grew a prosperous farm. In the early 1980s, they caught wind of a Corps plan to sell off some of its surplus land holdings they'd taken through eminent domain. The family bought a slice of their old farm back, planted some peach trees, and eventually hired Jones to look after the orchard. Twenty-two towns were lost to the dams, and some of the names make you wish you'd stopped by before they went under: Ainsworth, Almota, Ayer, Farrington, Grange City, Illia, Joso, Levey, Magallon, Matthews, Moore, Page, Penawawa, Riparia, The Riviera, Scott, Sheffler, Silcott, Simmons, Snake River Junction, Walker, and Wawawai. Thanks to one farming family with an indomitable loyalty to their land, one settlement has already been reestablished. The peaches along the banks of the Snake look like the seeds of a new Penawawa.

We launched in fetid slackwater thick with the summer's accumulation of aquatic weeds. A carp the size of a springer spaniel slunk among the undulating vegetation. We stuck close to shore at first so we could chat. I asked Jones about the farm. "We got by this year," he said. "Not a bad year but not as good as last year. It's still the same story. Not much room for the little guy. The line between solvency and bankruptcy keeps getting thinner. I worry that all that'll be out here in a few years are giant corporate farms."

He asked me about the murky world of salmon science and research. I told him how researchers had discovered the way salmon can recolonize territory from which they've been cut off. In the mountains of Mexico, scientists figured out that the trout in the streams are the descendants of steelhead who holed up high as their world warmed up and dried out. And on the mid-Columbia, they've observed that a solitary female steelhead that can't find a mate will be hit on by resident male trout as relentlessly as a frat boy with a mommy fetish will ogle a real housewife. The steelhead hen will eventually give in to the trout's advances. Their offspring will be anadromous.

"What about your year?" Jones inquired further. I told him about Butte Creek and the Kennebec and Alaska. "What's going on in these places is a sustainable kind of conservation or ongoing restoration," I said. "What's going on here is something that falls way short of any of that."

We separated and paddled on in windy silence. The wind pushed some two-foot swells on the normally calm surface of the impoundment. Playing on these waves, allowing the surge to surf the boats upstream, seemed to lighten the mind, open the door to brighter possibilities. A small-town, practical conservatism, a plain, well-scrubbed, and worn-in decency, evident in the likes of Fick and Jones, will be integral qualities of any long-lasting resolution of the Snake River's troubles. They share a sense that if the river becomes no good for salmon, it soon might not be good for much else. Each described a quality of small town life they see as endangered, a way of living they wished to preserve.

They're capable, I believe, of asking and answering the pointed questions that never quite get addressed in courtrooms

and boardrooms as the wealth of the river gets diverted elsewhere. What does the fisherman owe the farmer? What does
the data center owe the small town? What do we owe the
river? If these questions were answered honestly, it seems possible they might one day comprise a kind of Fick-Jones post-
industrial average, measuring the odds for a prosperous,
sustainable, peacefully coexisting farming and fishing segment
of an economy, moving, as *Cadillac Desert* author Marc Reisner
wrote it, forward to the past rather than backward to the future.

The breeze made quick work of the paddle to Penawawa.
We pulled our boats onshore across the railroad tracks and into
the orchard. I had been daydreaming all afternoon of the last
good peach of the season on some neglected tree high enough
to avoid harvest by deer but hidden enough to escape notice
from humans. But the demand for the peaches of Penawawa
was great enough that no stragglers were left behind. "Next
year," Jones promised. "You gotta come in August. The heat
we get down here then sweetens them, makes them just about
perfect."

"I will," I promised. "And the paddling today was fun. Riding the upstream waves. Maybe one to tell the grandkids about
after the dams come out."

"Well I'm ready. Ready for something different. Nothing
could be better," he confided, "than having an orchard along
the river."

Some people are called to farming, others to fishing, still
others to become intermittent and ineffective keepers of the
quiet, clean, and untrammeled places scattered among the out-
posts of civilization. There was no hesitating when the proposal
was made to make a final, late-season visit to the Buzzkill. With
that same loyal old friend who walked up the river with me
the month before, I flew in and floated out, saving the walking
for a mere ten-mile afternoon jaunt up the creek in question.
Launching in our boats onto a thin skin of river over rock, we
bumped down to the confluence with the Buzzkill, pushed on
shore, strapped on sandals, and lit out on foot. Not a mile up

the creek, we jumped three wolves from their daybeds. They looked as surprised to see us as we were to see them. I had envisioned wearing something other than a Hawaiian shirt and swim trunks for my first encounter with them.

Another mile or two up the creek, we rounded a bend to find a sow black bear and her two cubs splashing in the Buzzkill. The mom was poking her head deep into a few different pools as if she might be fishing. To say anything would have jinxed it, but we were starting to feel better about the odds of the lone salmon on the Buzzkill finding some company.

We saw a dozen salmon on the Buzzkill that afternoon, including one good-sized male, turned a brilliant crimson, that behaved as if he'd appointed himself to be in charge of spawning operations, a task for which he seemed well-suited. Seeing more than one fish gave me some sense of relief. But twelve is still far short of what the Buzzkill was meant to nurture. For the moment, however, twelve was a beautiful number. Having crisped ourselves in the sun on the granite cliff, we reluctantly headed down canyon back to the boats, stopping once to slake a considerable thirst in the cool water of the creek.

Camp that night was yet another broad white-sand beach, guarded over by ancient cedars, affirming the batholith's long symbiotic relationship to the Pacific. We toasted to coastal Idaho, the wolves, the whales, the bears, and the salmon. And it occurred to me a final metric might be added to the hypothetical Fick-Jones index. What is our responsibility to the fish?

Wise human voices have long recognized how our fate is bound to that of our fellow four-legged, feathered, and finned inhabitants of the earth. How they manage to make us feel this way, or why, is perhaps less important than honoring the long-held intuition that we won't live as well without them. That our future somehow seems tied to theirs eludes rational understanding the same way a young orca clowning around Puget Sound with a chinook balanced atop his head does. The punch line to these wordless jokes are worth endless deferral, the image more enticing than any possible explanation. It is a riddle worthy of all-out defense, even as the defenders recognize not everyone loves paddling upstream, poetry, and soul food.

Salmon can remind us of the urgent need to adjust the route on the direction home. They can inspire us not to relegate our grave concerns for the future to a pathetic nervous silent fretting over the slow rise of mean annual temperatures from Waterton to Walla Walla. Their absence for a portion of the year, and their much anticipated return, infers an existence in another part of the world that must be fertile, fecund, and healthy enough for them to get along there, keeping them strong enough to complete the journey home. This fosters a kind of faith that there are still many good places left on earth. Implicit to the development of this trust is the idea that one home cannot be damaged without compromising another, that the borders between places ought to be mapped as subtle, permeable, and fluid, rather than rail straight and impervious as concrete. Call it a generous sense of mutually assured creation.

Working back to a world bright with that kind of assurance requires a well-developed sense of pragmatism and realism. The effort will necessarily look quite different from place to place. And because it's not likely we'll soar too far beyond the patterns of behavior that have defined the human experiment, it's safe to say that consensus will be rare, and most every move will be hotly contested. There will be stark failures, irresolvable conflicts, irreconcilable differences, preventable tragedies, inconsolable losses. There will also be more than a few unlikely triumphs, stolen against some long odds. "A small group," as a Kennebec alewife warden said, "can make a tremendous difference."

Back on the river, we tired of proposing toasts and rolled out our bags beneath the stars. My compatriots took quickly to sawing logs. I lay awake listening to the water, dealing with a familiar bout of insomnia by working my way to the bottom of one of water's wordless riddles. A half moon rose and continued its nightly drift across its sea of stars, casting its reflected silvery light on the surface of the river, creating the same translucence as in a salmon's eyes. What world do they see through those quicksilver droplets? Like a grab at a fist full of water came the ephemeral touch of a reply. I felt the pull of the moon on the tide, saw the run of the river to the sea, sensed the edge of some equilibrium in the land between them, balanced as a still-hunting heron. Patient enough, a way would open up. A path to follow home again at last.

AFTERWORD

As *Recovering a Lost River* goes to press, the epic biological opinion court case on Columbia Basin salmon (see chapter 9) doesn't yet have a verdict. Regardless of what Judge Redden decides, further legal challenges will undoubtedly complicate any new federal biological opinion on salmon for years to come. But emerging alliances outside the courtroom have the potential to exercise at least as much influence over the future river as any litigation.

As I described in chapter 5, ExxonMobil, Harvest Energy, and ConocoPhillips are hoping to float huge pieces of Korean-made mining and refining equipment across the Pacific and up the Columbia and Snake rivers to Lewiston. The remainder of the 1,200-mile journey to the tar sands of Alberta would be made on roads. Massive trucks, 210 feet long, 30 feet high, and 25 feet wide, ferrying loads weighing up to 675,000 pounds—more than five times the federal standard for U.S. highways—would transform a route to northern Canada that would otherwise qualify as one of the world's great scenic drives, touring along the largest roadless areas in the Lower 48. One of those wildernesses is the greater Salmon-Selway, the last best stronghold for salmon in the western United States. Thus the fate of millions of salmon, and the future of one of America's best wild places, is being held hostage by Exxon, a South Korean steel company, and the unyielding desire on the part of boosters in Idaho to finally furnish Lewiston's port with an excuse for its existence. And that's not the half of it.

Cooking a barrel of oil out of Canadian sand produces three times the carbon output of a conventionally extracted barrel of oil. James Hansen, NASA's outspoken climatologist, has said there's no realistic chance to mitigate climate change if the tar sands are fully exploited. Merely cooking the sand to get the oil will spew the equivalent of 18 million cars' annual output of carbon into the atmosphere. In the meantime, tar-sands operations will destroy a boreal forest the size of Florida, kill millions of birds, poison the Athabasca River, and make Chipewayan and Cree people downstream sick.

In the wake of barges moving upstream on big oil's behalf, claims of dams' carbon-neutral virtue turn to little more than filthy bilgewater. In the absence of any meaningful recognition on the part of dam defenders that fossil fuels are becoming both scarce and a global health problem, dams are performing their usual dirty work: transferring wealth for the profit of far-off interests and, in this case, perpetuating fossil fuel dependence. Together, big dams and big oil are making rivers and roads in the Pacific Northwest accomplices to one of the worst environmental crimes of the new century. The legacy seems fitting: like the vast majority of dams built the century before, the tar sands will eventually cost far more than they ever contribute.

This "Grand Theft Hydro" pattern continues on other fronts. In October of 2010, the Wasco County district attorney, whose jurisdiction includes interpreting and enforcing the terms of Google's contractual gag order (see chapter 5), ordered local officials to reveal just how many people the data center was employing and what they were being paid. The DA found that the company, with the cooperation of compliant city officials, had been revealing even less than it had agreed to. At issue was whether Google was living up to the terms that granted it a fifteen-year holiday from paying local and state taxes. How much public hydropower Google gobbles remains a very private corporate secret.

Further down the Pacific Coast, for the fourth year in a row, despite relief from a long drought, Butte Creek spring chinook salmon were stranded by low-water conditions and had to be rescued from the lower reaches of the creek. Water diversions were taking 80 percent of the flow at the time of the rescue.

Finally, some good news: on a recent bright fall afternoon, I took the ten-minute drive from my house down to my home-town's namesake river, the Hood, to see it flowing free for the first time in eighty-seven years. The two-story-high Powerdale Dam, well beyond its useful life, sat four miles up from the confluence of the Hood and Columbia rivers, and blocked 140 miles of habitat for salmon and steelhead. In the summer of 2010, it was dismantled at the behest of the Confederated Tribes of the Warm Springs, the Oregon Department of Fish and Wildlife, and PacifiCorps, the utility company that owned it. At the site of the former dam, the riverbank and channel have been well reconstructed. The loss of the pool behind the dam revealed the river's picturesque confluence with Neal Creek, its susurrations audible for the first time in nearly a century. All that remained to be done was a couple acres' worth of plantings of native grass and trees. While I studied possible paddling routes through what looks to be a healthy class III or IV rapid flowing where the dam recently stood, one of the project's managers spotted a steelhead as it breached the surface in a near-shore eddy. I asked another manager, the man in charge of the deconstruction, if there were any complications to the removal. "None," he told me. "This is a small dam. We built our coffer dams to divert the river. We built a temporary bypass for the fish. Then we just started chipping away at the concrete." Would the work be any more complex, I asked, on a bigger dam? "No," he said. "If there's a blueprint for it, we can tear it down."

As he spoke, the possibilities of the future brought me back to the burning question that prompted me to write this book: where and how will my children live? Part of the time, I hope, near a good river somewhere, where they can wander down to the water with hook and line, boat and paddle, or just the shoes on their feet and find out for themselves what Thoreau meant when he said, "The life in us is like the life in the river." I hope the answer they come up with *is* like the river, resonating with grace and power. I hope they will talk, if they are so moved, as the owners of things that require no title, no fences, locks, gates, or doors.

And I hope there isn't anything like a monolithic dam, behemoth corporation, or oppressive government blocking their chosen path or their preferred river. But if there is, I hope they remember what that dam deconstructionist on the Hood told me. If there's a blueprint for it, we can tear it down.

Hood River, Oregon
October 13, 2010

ACKNOWLEDGMENTS

For help in getting me to say what I mean as well as mean what I say, I'm grateful to the sharp eyes and wise editorial voices of Chris Bouneff, Catherine Hawley, my agent Michael Snell, and Brad Tyer. Sarah Gilman at *High Country News* deserves credit for shaping an early draft of the science chapter. For her somehow laid-back, yet simultaneously intense line of questioning, suggesting, reordering, rethinking, rereading, contemplating, scrutinizing, and fine-toothed combing, to my editor at Beacon Press, Alexis Rizzuto—thanks again.

For insight into scientific matters, which are never strictly scientific matters, thanks to Jim Anderson, Robin Baird, Ken Balcomb, Bert Bowler, Ed Bowles, Ed Chaney, Don Chapman, Carl Christiansen, Michele DeHart, John Ferguson, Margaret Filardo, Greg Graham, Nate Gray, Steve Haeseker, Mike Kelly, Jim Lichatowich, Mark Lindgren, Bill Maslen, Michelle McClure, Bill McMillan, Fred Olney, Rock Peters, Rod Sando, Howard Schaller, Carl Schreck, and Rich Zabel.

Thanks to several scientists who took the trouble to talk to me on condition of anonymity. It's brave to speak when doing so puts your job on the line.

Comprehending the murky world of dam, salmon, water, energy politics, and the law is no mean feat. I owe gratitude in this endeavor to Dustin Aherin, Dave Bahr, Felicity Barringer, Michael C. Blumm, Karl Brooks, Reed Burkholder, Nicole Cordan, Dave Cummings, Angus Duncan, Brock Evans, Steve Fick, Jeremy Five Crows, Pat Ford, Greg Haller, Randy Hardy,

Allen Harthorn, Charles Hudson, David Johnson, Bryan Jones, John Kitzhaber, Sam Mace, Steve Mashuda, Rebecca Miles, Ken Olsen, Olney Patt Jr., Chris Shutes, Glen Spain, Jim Risch, Tom Dechert, Jim Kluss, Jeff Reardon, Todd True, Paul Van-Develder, George Viles, Steve Weiss, James Workman, and Eric Zachariason.

If I had to choose, I would still rather be a man out standing in his field than a man outstanding in his field. For their encouragement in keeping this priority straight throughout the research and writing of this book, providing the best kind of company and counsel over many miles of running, hiking, flying, fishing, and floating: thanks to Levi Carson, David Clarkson, Kevin Cook, John and Tricia Cox, Brian Eckart, Trevor Groves, Greg Hawley, Niels Maumenee, Carl Poston, Ben Sayler, Peter and Raluca Vandergrift, and Dick Walker.

Thanks to Al Jubitz, who was relentless in both his encouragement to write this book and his help on several fronts in making sure it happened, and to the late Mike Hawley, who also thought it was a good idea, with the caveat that if I was going to write a book about a fish, it had better be about a lot more than fish.

Thanks to Carol Hawley, for stepping in to help while I was away from home, but mostly for teaching patience.

Steve and Diane Pettit deserve much more than just their own sentence full of kudos, not only for giving me a crash course in Idaho "biolotics" but for taking time as well to patiently correct my misconceptions, answer a constant barrage of questions, feed me, tolerate my takeovers of Steve's office space and guest bed, and introduce me to several faces and places that made my research more fun and this book a much better read.

For the advice to add a voice to a prayer and a soulful note to a blues lament, and for all kinds of help in the composition of that song, my thanks to David Duncan.

Finally, to K.: years ago when I thought it couldn't be done, you were certain it could be. A thousand thank-yous.

NOTES

2: What They're Smoking in Alaska This Summer

1. In reality, these fish wind up traveling the opposite direction. Ninety percent of the commercial Alaskan salmon catch is bought by the Japanese.

2. These are all included in the genus *Oncorhyncus*, an appellation derived from the Russian term for "hooknose," which describes the adult male spawning salmon's hooked upper jaw. Two of the seven species are found only in Asian waters, *O. masou* and *O. rhodurus*. Five are native to Western Pacific waters: pink (*O. gorbuscha*), chum (*O. keta*), sockeye (*O. nerka*), coho (*O. kisutch*), and chinook (*O. tshawytscha*). Pacific trout within the genus include the Westslope cutthroat (*O. clarki*) and the anadromous steelhead (*O. mykiss*). There are regional names for each of these species. The chinook is also referred to as the king salmon, coho are called silvers, chum are called dogs, sockeye are known as bluebacks. The life cycle I describe applies to all species within the genus, with the exception of sockeye, who rear their young in high mountain lakes rather that swift shallow reaches of stream.

3: Feed Willy

1. Like unprotected sex, the pill has unintended consequences. The endocrine systems of creatures who live in or on water downstream from urban areas may well be altered by the concentration of hormone-disrupting chemicals in oral contraceptives. Potential side effects from other pills in this sewer-borne pharmacopeia have yet to be confirmed, much less imagined. God only knows, for example, what your father's boner meds will do to a bonefish.

4: Butte Creek

1. A grandson of archeologist Hiram Bingham, who uncovered the ruins at Machu Picchu, Northern California commercial fisherman Nat Bingham was pushing for dam removals on Butte Creek and

Battle Creek (another salmon-bearing tributary of the Sacramento) as early as the 1980s. A poignant description of the infectious enthusiasm Bingham shared for community-based salmon restoration is found in Freeman House's *Totem Salmon* (Boston: Beacon Press, 2000). Bingham died in 1998.

5: Energy Versus Eternal Delight

1. There's a vigorous global debate about whether dams should be granted mitigation points in some global system of carbon cap and trade. Research in Canada and South America shows some dams are concentrated sources of greenhouse gasses, especially methane, which traps heat in the atmosphere much more effectively than carbon. The hydropower industry, particularly Canada's Hydro-Québec, has tried to delay, squelch, or otherwise censor the results of studies on the matter. Dams that have been pegged as net carbon contributors have generally warmer or shallower reservoirs and lots of decomposing organic matter in their waters. The conventional wisdom is that the Columbia matches neither of these parameters. But water in these reservoirs has gotten warmer as summers have become longer and hotter. Fertilized by agricultural runoff, the warmer temps encourage massive annual blooms of aquatic plants. Water temperature in mid-August 2009, measured in the tailrace below Bonneville Dam, was 75 degrees.

2. Helping establish the precedent Google followed, Oregon's largest corporate utility, Portland General Electric, pays almost no taxes. The fiscal policy journalist David Cay Johnston has pointed out that Portland General Electric effectively charges its ratepayers for taxes it does not pay, amounting to a $900-million windfall from 1997 to 2006.

3. Hodel became Reagan's secretary of energy. A fundamentalist Christian right-winger in the mode of James Watt, in 2003 Hodel was hired as CEO of Focus on the Family, James Dobson's Christian media empire.

4. In 1998, former Avista Energy vice president of business development Judi Johansen became head of the BPA. Johansen was the right person to heed corporate utilities' complaint that payments from the BPA for the residential exchange program were too miserly. Like her predecessor, Hodel, Johansen signed on to a clever ruse that worked around a law to limit payments to corporate energy providers under the residential exchange provision, after which she accepted a job offer from one of the top beneficiaries of this newfound BPA largesse, PacifiCorp. In 2007, a federal court found that the scheme Johansen had helped cook up was illegal.

5. According to Taxpayers for Common Sense, the Bonneville Power Administration (BPA) has also claimed more than $500 million in credits on its debt payments for "salmon restoration" actions. The credits are issued whether any restoration occurs or not, as for example, in 2001, when the agency dumped water for power generation that had been earmarked for fish.

7: When the Levee Breaks

1. Marmes was the most famous but by no means the only significant archeological site along the lower Snake. At least eighty other sites were flooded by the waters rising behind the Lower Monumental Dam.
2. By my calculations, this translates to a minimum 135 standard-size dump-truck loads per day, 365 days a year—but going where? Along the Clark Fork River in Western Montana, a minor intrastate battle developed over where to put dredge tailings prior to the removal of the Milltown Dam just upstream of Missoula. As is usually the case, the side with the most people (and money) won. The highly contaminated tailings were shipped upstream to an outpost near Butte called Opportunity. There is not yet any analogous Opportunity for Snake River sediment.
3. The Corps gained ownership of all the waterfront property in Lewiston adjacent to the Snake by right of eminent domain. This represents quite a chunk of real estate, since what was coopted was the width of the old waterfront up to the waterline of the new reservoir, plus a buffer strip some eighty to one hundred feet wide.
4. The Port of Portland, in turn, also leans heavily on federal subsidy through the Corps. Portland sits about a hundred miles inland on the Columbia and Willamette rivers. To get there, a channel with a minimum depth of forty feet has to be maintained for the huge oceangoing tankers that work the world's marine ports. As ever-larger tankers are built, even with the maintained channel, these ships often leave Portland only half to three-quarters full to insure against stranding, a precautionary measure that creates considerable hassle and expense.

   The docks at Portland were quickly falling out of favor, dealing a potentially fatal blow to Lewiston's barges, until a decades-long lobbying effort by the respective ports and the Corps succeeded in procuring funding for a $150 million project to deepen the Astoria-to-Portland channel by three feet.
5. In the same e-mail, Feistner cited his own relationship to the Pacific Northwest Waterways Association (PNWA) and Doeringsfeld as a

means for Dechert to become more informed of the Corps' plans: "You might periodically talk with Dave Doeringsfeld at the Port since he attends PNWA events and I communicate often with PNWA," he advised Dechert. Indeed, Feistner is listed on the PNWA's Web site as someone who speaks ably and often to the lobbying group's constituents.

8: The Fifth H

1. Where the health of salmon are concerned in the Pacific Northwest, there may be no more fitting tribute to Gorton's career than the breaded whitefish filet. Quick to deploy legislative riders that mandated politically unpalatable anti-environmental actions, always in rabid opposition to Indian fishing rights, one of the last acts Gorton made in the Senate was a directive that would have prevented any federal money from being spent to study the feasibility of removing the four lower Snake dams.

2. The Corps couldn't possibly justify this move based on the content of that postcard:

> Dear NMFS and U.S. Army Corps,
>   As NMFS' draft science paper correctly finds, only dam removal can restore Snake Basin salmon to healthy numbers and fulfill our treaty obligations. The Corps exaggerates the costs of dam removal by ignoring the huge expense of dam retention and salmon extinction. Since these fish are already dangerously close to extinction, recovery efforts must begin immediately.
>   I ask that you recommend removal of the four lower Snake dams to save our endangered salmon. Please include my comments in the administrative record.
>   Sincerely,
>   From:
>   Name
>   Address
>   city/state/zip
>   phone
>   email

9: Lies, Dam Lies, and Statistics: The Science of Saving Big Hydro

1. With a nod to the sphere of influence enjoyed by the Columbia River hydrosytem, within the NOAA's six regions that cover American fishing waters, only the Northwest Region's directorship is a political appointment. All other regions have directors who are hired based on their qualifications as scientists. In June 2010, Will Stelle, head

of NOAA Northwest Region during the Clinton years, when a spate of bi-ops were found by courts not to meet the legal standard, was appointed by the Obama administration to take the job again.

2. In 2001, just before he was put in charge of salmon recovery, Walton testified in front of a congressional committee against dedicating spill for salmon and in favor of more hatcheries and barging.

3. Further information about pills poisoning rivers and streams can be found on the U.S. Environmental Protection Agency's Pharmaceuticals and Personal Care Products page, http://www.epa.gov/ppcp/faq.html, and on the Natural Resources Defense Council's Web site, http://www.nrdc.org/policy/documents.asp?topicid=60&tag=water+pollution.

4. The administrative record of the process behind the 2008 Federal Columbia River Power System Biological Opinion and the amendment to that document, the Adaptive Management Implementation Plan (AMIP), reveals more about Stier's direct influence over NOAA scientists. Wherever and whenever Stier could find what he interpreted as gray areas or gaps in any scientific data or analysis, he claimed these uncertainties as places where policy, not science, would dictate the standard for extinction risk. Identifying such gaps on at least one occasion involved selectively misinterpreting the work of actual scientists. In any case, uncertainty became a convenient excuse to further reduced standards that would dictate whether a specific population of salmon was flirting with extinction.

5. So just what is a "jeopardy standard?" Under the provisions of Endangered Species Act law, the U.S. Forest Service, U.S. Fish and Wildlife Service, and, where ocean creatures are concerned, NOAA-Fisheries must pass judgment on whether a proposed action within an agency's jurisdiction would cause further jeopardy to already listed endangered species there. In making this judgment, policy makers from these agencies must rely on "the best available science." Economic concerns are not to be taken into consideration. Yet far more often than not, the agency in question finds the action in question poses "no jeopardy." Contrary to the law, both politics and economic factors, wildlife advocates have pointed out, are usually lurking influences in reaching a "no jeopardy" finding.

10: A River Resuscitated

1. Federal dams like those on the Snake are exempted from FERC jurisdiction.

11: The Heart of the Monster

1. Stevens was not the first man to be made a fibber where tribal treaty negotiations were concerned. Of the 360 treaties the United States negotiated with North American Indian tribes, none has been faithfully executed according to the terms and conditions written into it. Many treaties were a fraud, and at least one, the Huntington Treaty of 1865, in which the Confederated Tribes of Warm Springs supposedly gave up traditional fishing grounds, was a total forgery. In 1997, Les McConnell, working for the Forest Service, published a report documenting irrefutable evidence of the crimes of the Indian agent J.W. Huntington for whom the 1865 treaty is named.

2. Fifty years after Grand Coulee Dam was completed, the Bureau of Rec finally compensated the Colville Reservation for the loss of lands inundated by the construction of that dam. Not to be outdone, in 2007, the Corps put the finishing touches on a housing project promised to Wyam people who'd lived at Celilo Falls—just in time for the fiftieth anniversary of the drowning of the falls.

3. While there's no smoking gun linking Gorton to the "Salmon Scam" fishing expedition, his fingerprints are all over the affair. As Washington's attorney general, Gorton led the fight to get his state out of having to comply with Boldt's 50 percent ruling. As Roberta Ulrich notes in *Empty Nets*, almost as soon as Gorton was sworn into the Senate, he was sponsoring an amendment to a fish-and-game law known as the Lacey Act, making a federal felony offense out of illegal fishing. Seven months prior to the amendment's approval, the Washington Department of Fish and Wildlife asked the NOAA to begin its stealthy investigation "in anticipation of changes to the Lacey Act."

4. From Ella E. Clark, ed., *Indian Legends of the Pacific Northwest* (Berkeley: University of California Press, 1953).

SOURCES

1: Redeeming the Dammed

Dams and Development: A New Framework for Decision Making (London and Sterling, VA: Earthscan Publications Ltd., 2000). The report can be downloaded at http://www.unep.org/dams/WCD/.

Jacque Leslie, *Deep Water: The Epic Struggle over Dams, Displaced People, and the Environment* (New York: Farrar, Straus and Giroux, 2005).

Dam Removal Success Stories: Restoring Rivers through Selective Removal of Dams that Don't Make Sense (American Rivers, Friends of the Earth, and Trout Unlimited, 1999). The report can be downloaded at http://sunsite2 .berkeley.edu:8088/cdri/search.execute.logic?simple=Dam+Removal+ Success+Stories+restoring+rivers+trout+unlimited&item=1.

John McPhee, *Encounters with the Archdruid* (New York: Farrar, Straus and Giroux, 1971).

The University of California at Berkeley maintains a Web database on the latest dam removals around the country at http://www.lib.berkeley.edu/ WRCA/CDRI/search.html.

Information on Sunbeam Dam came from the archives of the Idaho Department of Fish and Game, the Idaho State Historical Society, and the Land of Yankee Fork Interpretive Center in Challis, Idaho. The latter source houses a thick binder of oral histories recorded by Forest Service anthropologists and archeologists in the early 1970s. It includes the recollection of one Bill Sullivan, eighty-seven years old at the time he told his story, who has it that a night watchman by the name of Pete Ryan aided and abetted the saboteurs who blew up the dam.

Retired Idaho Fish and Game biologist Bert Bowler, considered by many to have the best grasp on the history of Sunbeam, thinks Idaho Fish and Game contracted to have the dam blown up. According to him, there is a record of payment made to parties unknown for dam deconstruction. Further clouding the matter, a portion of those proceeds was refunded to Fish and Game without explanation.

2: What They're Smoking in Alaska This Summer

The Alaska Department of Fish and Game answered all my queries about the subsistence fishing program, management practices, annual totals for the commercial catch, and the state's history of eschewing big dam projects.

Information on the natural history and life cycle of Pacific salmon:
Xanthippe Augerot, *Atlas of Pacific Salmon: The First Map-Based Status Assessment of Salmon in the North Pacific* (Berkeley: University of California Press, 2005).

Tom Jay and Brad Matsen, *Reaching Home: Pacific Salmon, Pacific People* (Portland, OR: Alaska Northwest Books, 1994).

James Lichatowich, *Salmon Without Rivers: A History of the Pacific Salmon Crisis* (Washington, D.C.: Island Press, 1999).

First Fish, First People: Salmon Tales of the North Pacific Rim, ed. Judith Roche and Meg McHutchison (Seattle: University of Washington Press, 1998).

Pacific Salmon and Wildlife: Ecological Contexts, Relationships, and Implications for Management (Olympia, WA: Washington Department of Fish and Wildlife, 2000).

3: Feed Willy

Interviews with Robin Baird (a whale biologist at Cascadia Research), Ken Balcomb (of the Center for Whale Research), and Candice Emmons (of the National Oceanic and Atmospheric Administration's Northwest Fisheries Science Center) provided me with a quick education on the southern resident's preference for chinook.

The risks to orcas posed by the continued existence of the Snake River dams was first brought to my attention by a cover article written by David Neiwert in the February 22, 2006, edition of *Seattle Weekly* ("Killer Dilemma").

John Ford and G.M. Ellis, *Prey Selection and Food Sharing by Fish-eating 'Resident' Killer Whales* (Orcinus orca) *in British Columbia*, Canadian Science Advisory Secretariat Research Document, 2005.

Historical abundance:
Chad C. Meengs and Robert T. Lackey, "Estimating the Size of Historical Oregon Salmon Runs," *Reviews in Fisheries Science*, vol. 13, issue 1, 2005.

Joseph E. Taylor III, *Making Salmon: An Environmental History of the Northwest Fisheries Crisis* (Seattle: University of Washington Press, 1999).

George Aguilar Sr., *When the River Ran Wild! Indian Traditions on the Mid-Columbia and the Warms Springs Reservation* (Seattle: University of Washington Press, 2005).

William D. Layman, *Native River: The Columbia Remembered* (Pullman: Washington State University Press, 2002).

4: Butte Creek

Marc Reisner, *Cadillac Desert: The American West and Its Disappearing Water* (New York: Viking Penguin, 1986).

Don Worster, *Rivers of Empire: Water, Aridity, and the Growth of the American West* (New York: Pantheon Books, 1985).

Chris Shutes of the California Sportfishing Protection Alliance spent several hours talking on the phone with me about the particulars of the Federal Energy Regulatory Commission's licensing of hydroelectric dams on Butte Creek.

Allen Harthorn and Friends of Butte Creek maintain a Web site, buttecreek.org, which updates developments with further restoration projects, fish counts, and ongoing negotiations with Pacific Gas and Electric to keep the water flowing.

5: Energy Versus Eternal Delight

Fuel leaks:
Scott Learn, "Fuel Barges Run Afoul on Columbia," *Oregonian*, July 21, 2009.

Snake River role in Kearl Oil Sands Project:
A series of articles in the Missoula, Montana, daily *Missoulian* written by reporter Kim Briggeman, beginning in fall 2009 and continuing as this book goes to press, outlines the issues of concerns to Idaho and Montana residents.

National Geographic, March 2009: Extracting tar sands' petroleum from bitumen "emits 3 times more carbon dioxide than ... [oil from] Saudi Arabia."

Google:
Ginger Strand, "Keyword: Evil," *Harper's Magazine*, March 2008.

Byron Beck, "Welcome to Googleville," *Willamette Week*, June 4, 2008.

History of the Bonneville Power Administration (BPA):
Richard White, *The Organic Machine: The Remaking of the Columbia River* (New York: Hill and Wang, 1996).

Blaine Harden, *A River Lost: The Life and Death of the Columbia* (New York: W. W. Norton & Company, 1997).

History of the Northwest Power Act (NPA) of 1980:
Michael C. Blumm, *Sacrificing the Salmon: A Legal and Policy History of the Decline of Columbia Basin Salmon* (Den Bosch, Netherlands: BookWorld Publications, 2002).

David Cay Johnston's analysis of Portland General Electric's tax dodge: "Fiscal Therapy," *Mother Jones*, January-February 2009.

Profits of Northwest aluminum plants during 2001:
Michael C. Blumm and Daniel J. Rohlf, "The BPA Power/Salmon Crisis: A Way Out," *Restoration: A Newsletter about Salmon, Coastal Watersheds, and People*, issue no. 26, 2001.

6: How the Mighty Were Felled

Karl Boyd Brooks, *Public Power, Private Dams: The Hells Canyon High Dam Controversy* (Seattle: University of Washington Press, 2006).

Tim Palmer, *Snake River: Window to the West* (Washington, D.C.: Island Press, 1991).

7: When the Levee Breaks

Keith C. Petersen, *River of Life, Channel of Death: Fish and Dams on the Lower Snake* (Oregon State University Press, 1995).

Carl Christiansen, Greg Graham, and Mark Lindgren of the Corps spent an afternoon with me discussing the problem of sediment buildup behind Lower Granite at the Corps' Walla Walla office in June 2008. With a public relations officer present in the room the entirety of the interview, the gist of the conversation was that the silt problem is "manageable" and that solutions would be readily apparent as the Corps' study on the issue is completed in 2012.

8: The Fifth H

"Comparative Survival Study (CSS) of PIT-Tagged Spring/Summer Chinook and Steelhead in the Columbia River Basin: Ten-year Retrospective Analyses Report" (Comparative Survival Study Committee and Fish Passage Center, 2007). The report can be downloaded at http://www.fpc.org/documents/CSS.html.

The National Research Council's remarks on the cost of water: *Upstream: Salmon and Society in the Pacific Northwest* (Washington, D.C.: National Academy Press, 1995). This NRC report also devotes much narrative space to the negative affect of hatchery fish on wild native salmon populations.

On this topic as well:
David R. Montgomery, *King of Fish: The Thousand-Year Run of Salmon* (Basic Books, 2003).

Michelle M. McClure, Fred M. Utter, Casey Baldwin, Richard W. Carmichael, Peter F. Hassemer, Phillip J. Howell, Paul Spruell, Thomas D. Cooney, Howard A. Schaller, and Charles E. Petrosky, "Evolutionary effects of alternative artificial propagation programs: implications for viability of endangered anadromous salmonids," *Evolutionary Applications*, vol. 1, issue 2, April 2008.

The Corps' conclusions on dam breaching can be found in "Improving Salmon Passage: Lower Snake River Juvenile Salmon Migration Feasibility Report"/Environmental Impact Statement (U.S. Army Corps of Engineers, 1999). The Corp states in this report that the Snake River dams were "designed with features to aid both juvenile and adult fish." In reality, as Chaney pointed out, none of the four dams was even considered for juvenile collection or bypass facilities at the time of construction.

The Environmental Protection Agency's comments on the Corps' plan to willingly violate Clean Water Act (CWA) standards were submitted on the EPA's behalf, April 27, 2000.

Whitelaw's debunking of the numbers in the Corps' analysis: "Breaching Dam Myths," *Oregon Quarterly*, Autumn 2000.

The series of *Washington Post* articles on the Corps was written by Michael Grunwald beginning in December 2000.

9: Lies, Dam Lies, and Statistics: The Science of Saving Big Hydro

The tension among the Fish Passage Center, NOAA, the BPA's private science consultants, and the BPA itself can be gleaned by looking at the memorandums published by the Fish Passage Center over the past several years, available for download on their Web site at http://www.fpc.org/documents/FPC_memos.html.

These memos are the format in which the Fish Passage Center responds to requests for data and information from anyone who asks. The responses are routinely vetted by NOAA scientists, as well as by private consultants working for the BPA or a lobbying concern sympathetic to the position of the BPA.

Kintama Research and the
Northwest Power and Conservation Council:
Welch's comments on ocean factors were made in a September 29, 2006, letter he wrote to Northwest Power and Conservation Council (NPCC) members, asking them to grant full funding for his new fish-tracking technology.

Anderson and BPA contracts:
Records of Anderson's payments—in fact, all of the BPA's outlays for Fish and Wildlife costs—can be found on the agency's Pisces software page, which tracks expenditures for the Fish and Wildlife program. Accessible to the public, instructions for using this service can be found on the BPA's Web site: http://www.efw.bpa.gov/contractors/usingpisces.aspx.

The BPA's Scott Simms confirmed Anderson's $100,000 payment in a February 2009 e-mail to me.

Lohn's comments on the Endangered Species Act were made at the November 14–15, 2006, meeting of the NPCC.

Documents obtained through the Freedom of Information Act shed light on the COMPASS controversy, the orchestration of Lubchenco's visit to Portland in May 2009, and the role of Stier in the creation of the 2008 bi-op.

Wright's comments to Northwest River Partners in December 2009 can be viewed at http://www.youtube.com/watch?v=79P251Z6lGA&feature=related.

10: A River Resuscitated

Elizabeth Grossman, *Watershed: The Undamming of America* (Berkeley, CA: Counterpoint, 2002).

Interviews with Viles, Lawrence, and Gray.

Jeff Reardon of Trout Unlimited gave his detailed account of the fight to tear out Edwards Dam.

Bates College economics professor Lynn Lewis created a buzz with her paper on the positive economic impacts of dam removal, using the Kennebec post-Edwards as a case study. Her paper, cowritten with Curtis Bohlen and Sarah Wilson, and originally published in *Contemporary Economic Policy* in 2008 (volume 26, issue 2), "Dams, Dam Removal and River Restoration: A Hedonic Property Value Analysis," can be found at http://pearl .maine.edu/windows/penobscot/pdfs/lewis_bohlen_wilson_CEP.pdf

11: The Heart of the Monster

Alvin M. Josephy Jr., *The Nez Perce Indians and the Opening of the Pacific Northwest* (New Haven, CT: Yale University Press, 1965).

Janis Johnson, "Saving the Salmon, Saving the People: Environmental Justice and Columbia River Tribal Literatures," *The Environmental Justice Reader: Politics, Poetics, and Pedagogy* (Tucson: University of Arizona Press, 2002).

Roberta Ulrich, *Empty Nets: Indians, Dams and the Columbia River* (Corvallis: Oregon State University Press, 1999).

Joseph C. Dupris, Kathleen S. Hill, and William H. Rodgers, *The Si'lailo Way: Indians, Salmon and Law on the Columbia River* (Durham, NC: Carolina Academic Press, 2006).

Dan Landeen and Allen Pinkham, *Salmon and His People: Fish & Fishing in Nez Perce Culture* (Lewiston, ID: Confluence Press, 1999).

Charles Hudson of the Columbia River Inter-Tribal Fish Commission (CRITFC) lent hours of his insight into the complex and shifting world of salmon politics within and among the four tribes represented by the CRITFC.

The recent history of the Nez Perce water rights adjudication was discussed in Michael C. Blumm's *Sacrificing the Salmon* (see also chapter 5 sources).

Epilogue: The River Why Not

Data gathered by the Interior Columbia Technical Recovery Team, outlining the favorable odds for extinction for wild salmon can be downloaded from the federal family's Web site, salmon recovery.gov.

Recent scientific study adds urgency to the considerable evidence that dams are killing wild salmon: Charles E. Petrosky and Howard A. Schaller, "Assessing Hydrosystem Influence on Delayed Mortality of Snake River Stream-Type Chinook Salmon," *North American Journal of Fisheries Management*, vol. 27, issue 3 (August 2007).

Phaedra Budy, Gary P. Thiede, Nick Bouwes, Charles E. Petrosky, and Howard A. Schaller, "Evidence Linking Delayed Mortality of Snake River Salmon to Their Earlier Hydrosystem Experience," *North American Journal of Fisheries Management*, vol. 22, issue 1 (February 2002).

Phaedra Budy and Howard A. Schaller, "Evaluating Tributary Restoration Potential for Pacific Salmon Recovery," *Ecological Applications*, vol. 17, issue 4 (June 2007).